DESIGNER'S NOTEBOOK

**A workbook
and reference
for the
designer**

DESIGNER'S NOTEBOOK

BILL BOCKUS

Pasadena City College

Macmillan Publishing Co., Inc.
NEW YORK

Collier Macmillan Publishers
LONDON

FRONT COVER
"Robot Head" Walt Mancini

Macmillan Publishing Co., Inc.
866 Third Avenue, New York, New York 10022

Collier Macmillan Canada, Ltd.

Library of Congress Cataloging in Publication Data

Bockus, H   William.
    Designer's notebook.

    1. Design, Industrial.   I. Title.
TS171.B6   745.2   76-4956
ISBN 0-02-311520-3

Change for the sake
of improvement?

Change for the sake
of change?

Change for the sake
of the planet?

to Jerry

who bought me my first
wind-up submarine, taught
me algebra and logic, and
let me win at tennis.

Acknowledgments

My appreciation and thanks to John Caldwell, John Dickerhoff, Larry Heliker, Charles Lewis, Mac McGavock, Don Munz and Bob Rahm for the many discussions and suggestions regarding three-dimensional design. Their knowledge of technical processes and classroom instruction as well as their professional experiences in the field have been invaluable. I am especially grateful to Randy and Marie Burrell for discussions and copy preparation in getting the whole thing together.

Four young men: Don Thomas, Bruce Brackman, Gregg Lewis, and Bill Bockus, Jr., have answered innumerable questions regarding the nitty-gritty details on hardware and mechanics besides giving their opinions on various processes and the relevancy of this information to beginning designers. Thanks again.

Preface

In today's world of throwaways, diminishing resources, and energy crises, the designer faces challenges he has never had to face before. How to make the most out of what we have. Can we use it over again? Is this robot necessary? Are questions that designers can really sink their teeth into.

Since the 1940's, designers and their education have become an increasing part of art schools, colleges and universities. Classes were actually called "Product Design" and along with them were introduced Graphics, Environmental Design, Interiors, City Planning and related engineering courses. Applied Design began to be recognized as a definite major in the visual arts curricula. As students began to understand the fact that there were job possibilities in these design areas, they began to take lower-division classes which would help them get some basic preparation for later work in the upper division (junior and senior years). As a result, the Applied Design Major classes have been gradually increased in the junior colleges, community colleges, private art schools and many of the smaller "feeder" schools that surround the university complex.

This text, "Designer's Notebook," then, is a collection of basic hardware, materials and processing principles mixed in with an overall, general look at what a designer is, what he does, and how he gets educated. It is written with the idea of letting young people know that industrial design is a profession, and that educational opportunities may usually be found somewhere near their home. Suggestions regarding projects and presentation procedures are included to help spark interest in the student working alone or perhaps provide a few ideas for the beginning teacher in the field.

This is a self-instructional type of text that includes comments, notes, and ideas regarding design jotted down over the years. It is designed to be used by the beginning art student. It lists the tools, design elements and production processes that concern a designer creating 3-dimensional products as well as those elements concerning a painter, sculptor, or printmaker.

Several problems are laid out in detail, step by step, to assist the beginning student in experiencing both flat and 3-dimensional visual design factors. Other, more functional, units with illustrated examples, are suggested to help the student prepare both presentation boards and actual models. These units contribute toward building a portfolio that can be of great help in entering an advanced school or finding employment.

The last section of the book contains information regarding the design field that has needed consolidation under one cover for some time: industrial design societies, national exhibit centers, industrial design schools, royalty contract suggestions and the professional designer's code of ethics are typical.

CONTENTS

SUPPLIES

General

1 pad 14X17 white layout bond
1 pad 14X17 tracing paper
1 T-square
1 graphite transfer sheet
1 X-acto knife - triangle blade
5 drawing pencils
   4H, 2H, HB, B, 2B
1 kneaded eraser
1 pink pearl eraser
1 roll 3/4-inch masking tape

PROJECT 1 Line 28

1 piece 6X12 illustration board
1 Rapidograph pen or HB pencil

PROJECT 2 Implied Line 29

1 piece 9X23 illustration board
1 sheet tracing paper
1 graphite transfer sheet
1 circle template
1 pencil 4H
1 Rapidograph pen or HB pencil

PROJECT 3 Flat Shapes 31

1 piece 10X14 white illus. board
1 piece each of black, brown,
  and gray paper
1 btl rubber cement
1 scissors

PROJECT 4 3-D Shapes - Paper 34

4 sheets 12X18 white construction paper
2 sheets 3 or 4 ply Strathmore
1 tube model airplane cement
1 scissors
2 pieces of scrap matboard or
  illustration board approx. 3X12
1 steel strip or steel rule approx.
  1 inch X 12 inches X 1/16 inch
1 hammer

PROJECT 5 3-D Shapes - Wood 38

1 Foto or drawing of transportation
  item as designated
1 sheet medium sandpaper
1 sheet fine sandpaper
  Scraps of light and dark woods.
  Approx. 1X1's, 1X2's or 1X3's
1 crosscut saw or band saw
1 metal hacksaw
2 C-clamps
1 tube model airplane cement
1 small can linseed oil
1 soft brush, approx 1/4" wide

PROJECT 6 Support Structure 40

1 interesting rock, six inches
  minimum dimension
  Several smooth wood scraps,
  various dimensions
1 screwdriver or hammer
1 btl glue or model cement
  Sandpaper
  Screws or nails
1 hand drill plus bits
  Linseed oil or other finish
1 rag

PROJECT 7 Rendering 42

1 Photo or drawing of machinery
  as designated
2 pieces 10X15 subdued color matboard
  Cutout textures from magazine scrap
1 scissors
1 btl rubber cement
1 set of paints
  Casein, acrylic or opaque watercolor
1 red sable brush, No. 6 pointed
1 palette or dish
1 water container
1 rag
1 mat knife

PROJECT 8 Oral Presentation 51

SECTION 1   THE DESIGNER

A.   What Is a Designer?

In a sense, anyone who creates something original is a designer.
Those who make something by copying or interpreting working
drawings usually are known as craftsmen, technicians, or artisans.

Those who make their living by designing products which are produced
in quantity are known as industrial designers. Buildings are often designed by a team of architects, structural engineers and interior
designers. Specialties like transportation, communications, or furniture provide a variety of work in a narrower band depending on the
individual designer's abilities and preferences.

A designer tries to make things better. "Things" can include any
part of the total culture of the world. Refusing to design an object
that was trivial or detrimental to the environment or society would
be as valid an example of a designer's philosophy as if he were
actually creating or improving a specific product. The designer is
a well-educated problem solver, and the net result he strives for is
to create a better total environment by his choices and actions. It
has been said that the artist should try to be the conscience of society.
If he doesn't lead — who will?

B. Where Does He Work?

When a designer leaves school he often works for a small group of associate designers who have banded together to defray administrative expenses like rent, telephone, secretarial help and drafting costs which each alone probably could not afford. He may very well do a certain amount of subordinate design, some drafting, answering the telephone, driving to the airport to pick up clients, dashing around to pick up sample of materials or getting component deliveries from vendors, etc.

These groups or associate offices often have a high rate of employee turnover as designers gain experience there and then go out on their own or to a particular client who may have liked their type of design work. Like other businesses, the small design offices tend to come and go and often fail for lack of sufficient starting capital or lack of prudent cost accounting. At any rate, when the student finally acquires experience he gradually gets established somewhere and may end up in one of the following situations:

1. Staff designer in a well-established industrial design firm doing:

   a. Exploratory research. Usually considered a long-range approach. The designer invents, uses new materials, discovers new uses for old materials, plans for future sales possibilities.

   b. Product development and testing. Keeps contact with exploratory research. Works with engineers and technicians on current products in production. Tries to balance the traditionalist's fear of the unfamiliar with new ideas. In the area of product development there is often a closely knit team which works together, interacting with each other's ideas. For example, one type of team could be composed of 3 members:

      • A designer concerned with form, color, texture, materials, proportion, processes, and function.

      • A mechanical engineer concerned with properties of materials, stress, heat, die cutting, and processes.

      • A production foreman concerned with fabrication costs, assembly procedures, and getting it all together economically.

2. Working with a manufacturing company designing their line of products, such as a toy company, tool company, swimming pool equipment company, and also acting as liaison if outside designers are employed.

3. Employed by a firm which makes or distributes a certain material (such as expanded metal or paper) to act as specialist in adapting the particular material to clients' needs.

4. Designer in an advertising agency. Agencies today often design exhibits, signs, store fronts, interior displays and the many other three-dimensional items like paper sculpture or packaging that are typical of graphic communications. The designer often designs, or at least helps supervise, the production of catalogs, assembly procedures, operating instructions or field maintenance manuals. He is often consulted regarding advertisements or advertising campaigns also. There is nothing that can lose a company its customers faster than advertising the wrong specification on a product or claiming it can do something it can't. Thus, the marketing or sales people tend to work closely with the designer, who is very apt to know more about a particular product than anyone else.

5. Self-employed. Accepting design jobs in a variety of areas as consultant or research agent is known as freelancing.

6. Working as consultant to wholesalers or retailers in purchasing areas. Mail-order houses often retain designers on their staff to help design or select and test items. Firms making or distributing sporting goods, automobile accessories, and hospital equipment, for example, require knowhow that is not always part of the background of a straight purchasing agent. The consultant-designer can often be more objective than the inplant designer. Firms use lawyers, ad agencies, auditors, air conditioning engineers, structural or accoustical engineers as outside sources — — why not outside design help as well?

7. Working for government or other agencies in selection, research, or testing areas.

8. Magazines sometimes retain designers in various fields, particularly in consumer research.

9. Teaching product design full or part time in public or private schools.

3

Within these work areas he may have special qualities that would make him a successful:

☐ Manager
An organizer with the ability to encourage people's use of their own strengths. Arbitrator. Central control.

☐ Consultant
A conceptualizer, a statement maker, thought provoker, idea provider, brainstormer and modifier.

☐ Artisan
Skilled technician. Knows what could work and suspects what won't. The feet-on-the-ground, nuts-and-bolts guy who has to make certain the thing goes together at the end of the line. Question asker. He can be as creative as anyone by showing another direction or fork in the road to achieve the objective.

☐ Expediter
Knows fabrication processes, material and component sources. Provides substitute processes and materials. Pinch hitter. Untiring and inventive. Often an ex-purchasing agent.

☐ Teacher
Recognizes principles of the design profession. Is able to Synthesize these and impart them to students with enthusiasm. Develops curiosity in his pupils.

☐ Homemaker (Do-it-yourselfer)
This area does not fall under any particular industrial or salary category, but it can make your income go a lot further. The guy who designs and builds his own house and furniture, or sailboat, or brick walk, or planter, or swing, or what have you, is probably one of the happiest designers alive. He conceives, he plans, he redesigns, he expedites, he teaches his children, he delights his wife, he builds with tender loving care and is totally involved in a variety of problems all the way. What a way to go! I know. I do it. Below are a few examples my family has tackled. They range from refinishing bentwood chairs and laying 9000 bricks on sand in patio and driveway to outdoor lighting, to tables, to garden chairs, to desks, to a potter's wheel, a 4 x 8 foot dollhouse, a play car made from an old wood filing cabinet, a chess set, a mini-bike, a go-cart, tennis court for $400, 15 x 30 swimming pool, studio, foldout davenport, mountain cabin and the last monster was a 12-foot outboard hydroplane designed from scratch — — material cost $85.

Designing the above items has been as satisfying to me as doing layout and design work for engineers working on 85-foot dish radio telescopes and dynamic measuring instrumentation, or working on presentations of a radio plane drone for anti-aircraft practice or preparing confidential operating instructions on a laser communication system for a moon probe.

When things become very large or complicated it is difficult for the young designer to remain personally or totally involved. It is hard for him to feel that the tenth doorknob he has designed for the skyscraper office buildings being produced by an architectural firm is a very important contribution. Well, we say, he has to start somewhere . . . . Yes, but unless he is encouraged and told that there will eventually be more significant work fed to him he will undoubtedly "split" and try elsewhere.

This feeling of not being particularly "involved" in the total concept probably accounts for the rapid turnover of young designers more than any other factor, including salary. Encouragement, warmth, and employee recognition are damn hard for most businesses to achieve. Head designers that have the ability to take into consideration "Human Engineering" in their own department as well as in their own product undoubtedly have the better design teams.

Another point a young designer has to keep under consideration is creative design versus administrative design. The higher one goes up the managerial ladder the further one gets from the drawing board and design decisions. Some men enjoy the management challenge, but they too sometimes become jaded and wish they could shuck the eternal shuffle, shuffle, shuffle of papers and return to the drawing table. Design at the top level of industry is sort of a hazy organizing job of using vast resources to achieve an end product. Exactly where the designer becomes only a businessman is a mushy gray line every designer must define for himself as he goes "up?" the ladder.

After you have gained the necessities of life, look around a bit before you go after the "larger" salary or up the management ladder. Salary is not always all money. You may discover that released time can be as important as money during middle age. The "larger" salary sometimes means giving an excessive amount of time to industry and having no time for living. Possibly by staying at your optimum design level you can use some of the extra time for playing or camping with your kids, creating other works of art, writing, or just looking at nature, instead of getting MORE and BIGGER. In other words, "Beware!" To know yourself is an insight more young men could well use at the age of thirty. If you like sales, the business-organizing paper-work end of designing or management — more power to you. BUT, if you like drawing, design formulation and execution, don't leave it too far behind.

ʃ

C.   What Does the Designer Design?

1.   Packaging
Can vary from a simple rectangular candy box to a plastic
shell for a cake mixer motor, a glass dispenser bottle, a gunny
sack, or the tank on a railroad car.

2.   Styling
Often refers to redoing or originating the "skin" or cover for
a piece of equipment which has already been engineered.

3.   Exhibit Design
Trade fairs, expositions, models, merchandising racks,
displays.

4.   Interior Design
Furniture, lighting fixtures, hospital special equipment,
restaurants, office equipment, airplane cabins, ships' lounges,
homes.

5.   Architectural Units
Signs, symbols, sculpture, buildings, store fronts, beach
fronts, doorways, playground equipment, and sometimes
related landscaping.

6.   Consumer Products
These would include appliances, hot water heaters, bicycles,
toasters, toys, lawn mowers, etc. Items that can be repaired
by semiskilled craftsmen.

Designs for the handicapped.

 The following six areas can rarely be handled without structural,
mechanical and electrical engineering involvement. The finished
product can be repaired only by highly skilled craftsmen.

7.   Transportation
Motorcycles, autos, buses, mail trucks, boats, dirt-moving
equipment, trains, etc.

8.   Industrial Equipment
Machinery and controls of all kinds. Measuring instrumentation such as thermostats, accelerometers, pressure devices.

9.   Agriculture Equipment
More of an industrial engineering problem.

10. Military
Weapons of all kinds, communications, search gear, etc.

11. Space
Human engineering is another term for best fitting the environment, the controls, and readout units to the human form and capabilities. As the engineer builds he must work hand in hand with the designer and the medical men to insure optimum results.

Obviously the larger and more complex a design becomes the more it requires a team effort. No one man can hope to have all the necessary experience within his short life-span. Pooling total resources and having complete cooperation between the men involved is the only way possible to design computer systems, spacecraft and starships.

The areas above overlap, and often, as the job increases in size, it requires more and more engineering information. And if you feel there is plenty of design work to be done on our own planet in relation to humanity and ecology before we build starships -- then don't work on starships. That * certainly is your prerogative.

12. Earth Design

This is a new field; or at least a professional recognition and organization of factors affecting the total design of the earth. For example, population density, food, shelter, and products are inventoried, plotted, and studied in relation to psychology, cultures, energy networks and world transportation. Solutions regarding problems of transportation, unplanned logistics, and pollution are then formulated and presented to all nations through radio, newspapers, magazines, and TV networks. Suggestions regarding design work to benefit humanity are part of the parcel. (See "World Game" end of Chapter 10.)

* If you are critical of money spent on Space, don't forget that, altho our present budget on space is about 3 billion (1.4 cents out of tax dollar) we also GIVE away almost that much in foriegn aid each year. And that, as individuals, we spend 17 billion per year on tobacco & cosmetics. Check also on farm subsidies which keep the inefficient farmer in business. Check your statistics first, ... then list your priorities.

D.   Where Can the Designer Get an Education?

Section 10, General Information, contains a list of schools teaching Product Design or related three-dimensional design programs. The majority of the large state universities have similar programs, but to save space they were not included in the list. Checking this list may pinpoint a few schools near you.

It is difficult to classify or to even point out strengths of various schools, as they often change their emphasis from year to year. It is much better for the individual student to write to the registrar of some of these schools for a catalog or talk with instructors from industrial design oriented schools near his home. Then, if possible, visit the campus during a school day, rather than just a weekend when everything is closed and vacant. Sometimes it is also possible to see schools farther from home, if you take vacation trips. Or you might know someone on the other end of the state or country that would drop in on the campus in question to see what the classes offer and write you about them.

Catalogs of schools often give you a good overall glance at the emphasis a school has on a certain major. If the catalog has a variety of classes in product design, rendering, advanced perspective, environmental design, production engineering, plastics, etc., you can be pretty positive it has a going program in the industrial design areas. Granted you may not be able to afford or get in the "best" schools, but my contention has always been that the attitude and drive of a particular student is the catalyst that makes a good designer. A school usually has ten times the amount of information and environment that even the talented student could begin to assimilate in a few short semesters.

So get to the best school you can and give it all you've got. If necessary you can do outside problems, you know. You don't have to coast because others do. And even in the most limited schools, you almost always find a few enthusiastic gems of instructors that will help you educate yourself. A sage once said something to the effect that "If you really want to educate yourself, find an important problem and set out to solve it!" Some pupils I know have even gone to a college with the idea of helping make it a better design school by their own efforts! How about that approach?   Have confidence in yourself.

!

CURRICULUM

The following class titles and sequence are culled from a number of design programs throughout the United States. This core program is meant to be helpful to a beginning student in identifying the many kinds of classes related to a designer's education. It is not necessarily presented as a model curriculum for a four-year college.

TYPICAL PRODUCT DESIGN CORE PROGRAM

(The 72 elective units would include the Science & Humanities requirements.)

FIRST YEAR	1st Q	2nd Q	3rd Q
Art History	3	3	3
Two-Dimensional Design	3		
Beginning Drawing	3		
Lettering	2		
Electives	4		
Three-Dimensional Design		3	
Beginning Painting		3	
Perspective & Rendering		2	2
Electives		4	
Beginning Life Drawing			3
Product Design (A)			2
Model Making (A)			2
Electives			3
	15	15	15

SECOND YEAR			
History of Design	3		
Product Design (B)	2		
Industrial Materials	3		
Model Making (B)	2		
Electives	5		
Industrial Processes		3	
Exploring Woods		2	
Beginning Photography		2	
Rapid Visualization		2	
Electives		6	
Interior Design (A)			3
Product Design (C)			2
Exploring Metals			2
Intermediate Photography			2
Electives			6
	15	15	15

THIRD YEAR	1st Q	2nd Q	3rd Q
Industrial Design (A)	3		
Graphic Design (Art & Copy Preparation)	3		
Exhibit Design	1		
Exploring Plastics	2		
Electives	6		
Design Methodology (Problem Solving)		3	
Interior Design (B)		3	
Printing Processes		2	
Electives		7	
Industrial Design (B)			3
Packaging			2
Advanced Rendering			2
Textile Design			2
Electives			6
	15	15	15

FOURTH YEAR	1st Q	2nd Q	3rd Q
Advanced Industrial Design (Traffic, Transportation, Shelter, Mass Production)	3		
Specialized Structures	3		
Transportation	3		
Environmental Design	3		
Electives	3		
Special Studies, Ind. Des.		4	
Electives		11	
Special Studies, Ind. Des.			4
Electives			11
	15	15	15

Recommended Art Electives

Illustration
Sculpture
Ceramics
Metalsmithing
Architectual Drafting
Cinematography

E.   What are His Rewards?

The immediate success of an industrial designer is usually
measured in terms of sales and his commissions or salary.
However, there are other aspects besides the product which
are difficult to measure but certainly are as important in
the long run.

The designer's ethics, his commitment to the client and the
ecology, his interest in his profession and his attitudes toward
helping young people are all part of a man's character that
remain long after financial gains are forgotten.

Today moral commitment must become more important than
technological achievement. Material wealth, alone, has
never been insurance for a moral society. Before you design,
examine the morality of your action. Do you want to use up
the resources of the earth for an automatic toothbrush? A
smog producing car? An eye-searing blinking billboard?
Another bulldozer? More hair-spray cans? At least start
examining your objectives. If your background is extensive,
if your education has been broad, if your commitment to
humanity is solid, you will probably come up with intelligent
sensitive answers and you will not have that spider* of
ecological guilt crawling around in the back of your mind.
Furthermore, your work will be done with enthusiasm and joy,
and the results will reflect your commitment and character.

And, of course, the greatest reward of all is seeing something
created that has sprung from your own mind. This joy of self-
expression is one of the greatest feelings in the world. All
the worries, the midnight oil, the tensions that build up while
one is searching for a solution and putting the whole thing
together seem suddenly to be worth it when that first product
comes off the line. It is akin to a sense of relief, a part of
total relaxation. Often the designer sits and looks at the
item for long periods of time as if it were a painting or sculpture until he gets his fill of "admiring himself." But then a
few days go by, and all of a sudden ideas begin to crop up
again, the next problem starts forming, a sort of tension or
challenge builds up, and "HERE WE GO AGAIN!" It's
really a great life.

*Apologies to the good spiders in the ecology.

2 CONCEPT & APPROACH

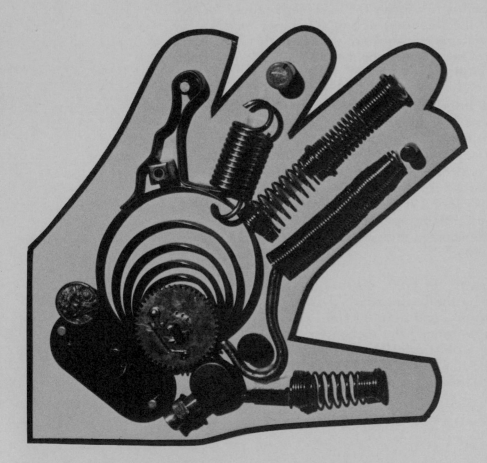

DAWSON

SECTION 2   CONCEPT AND APPROACH

A.   Introduction

Concept is the designer's analysis of those factors which must be satisfied to achieve a solution. It is his visualization of the finished product.

Approach is the manufacturing process and materials used to build the product.

While most of us are loud in lauding our design achievements in industrial machines, architecture, and transportation, it is rather interesting that when we "do" our own home we find the greatest satisfaction in design from yesterday. The 1850 Chickering box grand; a 100-year-old condiment castor; several bentwood or sturdy captains' chairs; a pair of two-foot-high oak candlesticks; a colonial high-back rocker; plus a well-padded, massive davenport; a grandfather's clock; and two overstuffed armchairs are very likely to be the core of our contemporary living-dining room area.

As a young man fresh from college I used to decry this philosophy of taste. "Why not use our contemporary designs? Be true to your own time!" I would argue. Thirty years later I find I am discussing the same thing with student designers who seem to have the same fervor I had and the same standards for "honest use of materials," "the home a machine-for-living," "contemporary use for contemporary materials," or whatever the current slogan is to express the designer's regard for the environment for the sleeping, eating, cooking, reading, TV viewing, entertaining homeowner. Yet today's living rooms strangely enough reflect just about the same design philosophy as they did in 1930.

In talking to an interior designer a while back I asked him what his guess was about the amount of period furniture sold in relation to the more contemporary designs. He answered, "I don't have to guess, man. Look in any furniture mart or outlet. Mediterranean is strong at present with about 40-45% of the market. Chinese Modern is almost passe', French Provincial is still around, Colonial is holding its own at about 25% of the market, and Contemporary is a bare 5% of sales, with the majority of Contemporary sales being to restaurants, offices, or industrial plants."

As the saying goes, "In 1900, contemporary design entered the American home by the back door." That is, the gas-electric stoves, refrigerators, and stainless steel sinks replaced the wood-burning, cast-iron range, the icebox, and the porcelain sink. The lavatory, the plastic toilet seat, and the shower replaced the outhouse and the bath tub. Great. But the trend never seemed to get to the living room. Why?

This is a chapter on Concept & Approach. The above paragraphs are one example of a definite gap between the needs and desires of the buying homeowner and the contemporary designer's concept and approach. Somewhere along the line the concept of what people want and need in the way of contemporary furniture accessories for their living rooms and bedrooms has faltered. And if even part of the designer's concept was basically correct, then the second step, the approach, was evidently wrong enough to destroy any possible validity of the original design concept. Let us take one example—an occasional chair. "Occasional" implies that it is used to sit in occasionally and that it can be moved easily to accommodate visitors. (Somewhat different would be the design concept of a "Master" chair, which is usually larger, heavier, stays in one spot, and is used by the family members to flop in or read in for extended periods.)

The contemporary chair very often has a number of advantages: It is light, structurally strong, and made of materials which lend themselves to mass production, such as wood, aluminum, steel, plastics, and synthetic fibers. It is reasonably priced. Why don't more people put it in their living room?

After design discussions I asked several classes to react to photographs of chairs of Early American design and to chairs of contemporary design. The reactions are listed below:

Early American	Contemporary
Handmade	Plain
Irregular surfaces	Smooth
Used	New
Compact	Graphic
Antique	Inexpensive
Comfortable	Angular
Warm	Cold
Strong	Weak

Obviously this is not a very comprehensive market survey by any stretch of the imagination. But there is enough meat here for a good class discussion when one realizes these words were repeated by three different classes which were unaware of each other's reactions.

Note that the modern "look" is not listed as uncomfortable. However, it was not listed as comfortable either. A reaction in some instances which seems to be an advantage may actually be a marketing disadvantage. For example, the word, Graphic, implied to the students that the silhouette or form of the modern chair was often more interesting to look at, say as a piece of sculpture, than its more compact "Early American" rival. "Fine," we say, "Boy, our market surveys show that people prefer our new chair's form." As we well know, what people say they prefer in a survey and what they buy are often unconscious opposites or, more likely, their statements are mis-interpreted by the survey analysts. It may be that a room full of sculptural "graphic" chairs would not relate at all to the homeowner's idea of a relaxed atmosphere. Even one modern chair — the one that looks so beautiful in "Our Home" magazine standing alone on the beach at sunset or poised on top of a sand dune, may or may not fit into a person's living room. Getting a charge out of the visual form of the chair is one thing. Using that same emotion as a basis for purchasing the chairs that one lives with all day long may well be another.

It is also possible for a chair to be structurally strong but look weak. For thousands of years people have associated solidity and strength with heavy timbers, rock and the post and lintel type of structure. Altho the cantilevered steel, the thin tubular extrusions, and angle-iron construction of contemporary chairs give all the necessary strength to the chair, the general appearance of these thinner members is evidently not convincing to modern woman. The chair is strong but appears weak. It may take another thousand years of evolution to change these primitive traditional feelings. In the meantime the slim, sculptural, strong, contemporary chair remains a great photograph, possibly a part of the dinette set in the breakfast nook but doesn't quite make it to the living room.

In the approach to the living room problem, uncovered aluminum or steel didn't seem to please Mrs. Housewife either. Exposed wood and fabric are still the top choices, probably because they seem warmer and softer and more conducive to relaxing. In patio or garden occasional furniture, these disadvantages remain. But the homeowner, in this case, places utility ahead of comfort and relatedness. Leaving furniture outside is a rather recent idea, and the light folding aluminum armchair with the multicolored plastic webbing can take a beating from the sun and rain that wood and fabric could not begin to challenge. In the case of outdoor chairs we might say then that Utility seems to be the design CONCEPT that is most important, and it is in this area that contemporary materials may be able to hold their own.

At present, softer vinyl arm rests, wooden back slats and even wicker weaves are appearing on outdoor chairs. Putting a bare forearm on a metal arm rest that has been sitting in the sun for 2 hours is not the most pleasant sensation in the world. Teflon-like materials may offer a more comfortable temperature to the sitter than bare metals, but it remains to be seen whether these plastic dip or spray coatings can endure expansion and contraction without cracking or separating from the underlying material.

If feasible, the use of these plastic coatings will probably make possible the development of better furniture. It would allow quiet colors, less reflection, no oxidation, better wearing qualities, less heat transfer and a warmer, softer feeling than metals. Eventually a warmer contemporary design may make it through the front door into the living room, as Colonial, Mediterranean, and French Provincial designs mush together and leave by the side door.

B.   Definition

The preceding paragraphs may give the student reader some idea of what is meant by Concept and Approach. The two are intertwined, but in general, I suppose a design concept implies a formulation of ideas which defines your objective. Then the approach, or production methods you apply to reach your design objective, will tend to jell or modify your original concept as you proceed. An approach might precede the concept or at least spark it. There are plenty of examples where men, fussing around with processes or materials, got an idea for an end product. For example, there is the story about a man fooling around and bending some old piano wire (a very stiff, thin, shiny wire) until he happened upon the idea of a stapling machine. He drew up his design, got a patent and started marketing the staplers. Another interesting item tells of two young men who were stretching some nylon sheets over a rack and got the idea of using taut nylon sheets as tent-like walls in an international exhibit. The lights were placed behind the nylon so that the exhibit pieces were bathed in a soft, nonshadow-like glow. In both these cases the designers used a variety of approaches (manipulation of a finished material) to spark or originate an original concept. However, even though manipulation sparked the concept, there was plenty of work still left to be done on defining the concept and channeling the approach.

Can we say in conclusion then, that the designer's concept of what constituted a well-designed occasional chair for the contemporary living room was, in many cases, not completely valid, because the importance of warmth, comfort and relatedness to the home were subordinated to lightness, graphic image, and contemporary materials? That the visual approach of thin structural members, exposed metal, and irregular silhouette gave the impression of weakness? And that the two together have been the wrong market solution up to this time?

Can we say the designer's concept of what constituted a well designed patio chair was more valid or that he happened to "luck out"? Was the need for utility (exposure to the elements or folding) recognized as the most important part of the concept? At any rate, the contemporary materials (light metals, thin members, plastic webbing) provided the correct approach, and the Contemporary furniture filled the need for the outdoor patio just as the Colonial and Mediterranean still seem to fill the need in the living room.

C.   The Spark

When some unique product appears on the market how often have you heard someone say, "Why, I thought of that years ago!" in the sort of tone that implies he would have made thousands if he'd only produced it. As any experienced designer knows, there are creative ideas occuring to people all over the world, in all countries, among all races and at every economic level. There have been cases of men finding a new breakthrough in science or medicine, for example, in two different countries within days or weeks of each other without even knowing the other was even alive. However, an idea is one thing, but the faith to risk money and effort producing a result is another. Very, very few people have the conviction, the desire and the drive to follow through on a creative thought. The great majority of people tend to "stick to their own potatoes" and leave the inventing or design risk taking to "those who know what they're doing"—namely, foundations or corporations with plenty of back up funds to support the production and research that do not pay off in dollars and cents.

And these individuals are probably right. But leisure time has led many of these same people into the arts seeking an outlet for their creative ideas in ways where they don't have to risk the family bank account. Getting into product design, for example, can be one of the most satisfying (albeit least known) ways to consummate that creative urge if you are interested in the arts. Creating ideas and useful or better products for society or yourself can be just as satisfying as oil painting, ceramics, or printmaking. At any rate, here you are, starting as a designer in the three-dimensional area. Your purpose is to find functional solutions for your own or society's needs; integrate these solutions with better form; and do it in a manner that causes the least disturbance to ecology.

A large order. And you'll need all the background experience you can get to follow through. Courses in Drawing, Design, Product Design, Drafting, Biology, Physics, Ecology, Welding, Chemistry, Sculpture, Weaving, Sociology, Crafts, Architecture, Advertising Design, Plastics, Interior Design, Psychology, etc. can all be pertinent.

"But," says the high school student or college freshman, "I haven't had all this background yet. How do I find a solution to a given problem in these early courses?" Part of the answer is "Think hard and long." And within your limitations you will find a reasonable answer. The idea of thinking "hard" is difficult to put across to many students, and yet is the only way solutions or inventions are ever discovered.

A great solution to a design problem is usually found by being obsessed with it. Most creative men admit that when they are on the trail of something they think about it morning, noon and night. They lean out of bed at 3 a.m. to scribble notes on it. They design during movies, while shaving, and in the shower. The obsession is there, not as anxiety but as a challenge or game. If the beginning designer will cultivate the habit of thinking hard and long on a problem, making notes, drawing quick sketches, coming back to it time and again, turning it over and over, trying a dead end two or three times more, until he is certain there is no gold in that alley, etc., he will discover that his brain will do miraculous things for him. Suddenly solutions occur, often two or more at a time. These flashes of comprehension don't always come while you are doing the hard thinking. They sometimes reveal themselves later, when you least expect it, during intervals when you aren't particularly thinking about anything. It is my opinion that thinking hard and long about a design problem often tends to simplify it. At least it seems to reduce the various parts of the problem into categories and tends to find relationships of the different facets. Somehow, then, after the problem has been "stated" clearly, this "statement" acts in the brain like a program for a computer. And even though you stop thinking about it, or go to a party, or go to sleep, it seems as if the brain keeps on organizing, switching synapses, and moving little "box cars" or bits of information around until a point is reached in your subconscious where the program is fulfilled. Then the related necessary information or solution surfaces to your more conscious channels and, Voila', you have an answer. It may not always be the final one you use. It may only be part of the answer. And it often comes while you are thinking hard too, or sketching. But one thing is certain. No good design solution was EVER found by producing the first shallow ideas sketched out on the drawing board. If, at the very least, you make certain of a clear concept, and a feasible, economical, approach, you have half the battle won. The details, the expediting, the jigs, the costs will all be that much easier to solve, if your objective has been carefully researched and thought out.

The most ingenious approach or production process, the hardest work in the world, the cheapest price have never produced a good product from a weak concept.

D. Considerations

As a designer of consumer products then, it might help to ask yourself the following questions as you gradually define your concept:

1. Will the product relate to the user and his surroundings? Or does the product only call for attention with little regard for appeal? Does it make the person using it feel as if he belongs in the same environment as the object? Or is it so different, so way out, so tradition-less that it alienates or repulses?

2. Does it function emotionally and visually for the user, as well as physically? Or is it designed for other designers to "Oh" and "Ah" over?

3. Is it your intention to design a product that will last? A lifetime? Or will it be trivial junk to be cast aside in a matter of weeks - another spoiler of earth's raw materials?

4. Will it enhance the environment? Or is it another "well-designed" billboard?

If you come up with mostly "Or's" in the above quiz, you will probably find future sales slow and your career in industrial design limited.

The moral or warning to you, the neophyte designer, is that concepts should be looked at very critically. Designers are human, and ordinarily they hate to be told, "It won't work." What you should do is to inquire. "OK, It won't work or sell . . . : Why?" Then get all the criticism you can from quali-fied sources and sift the destructive as well as the constructive criticism.

E.   Critiques

In a design class it is often quite illuminating for the student to present his "concept" to the class with a few strong sketches before he starts his "approach" thinking. Let the other students criticize his "concept" ideas both pro and con. Discussion might cover need, use, market, competition, esthetics, psychology and ecological implications long before it touches on engineering production problems.

There is a great difference between an item produced to fit a complex machine need and an item produced to satisfy a consumer's esthetic need. The coil spring for an automobile front wheel is an item that is designed by an engineer, and there is little emphasis on esthetics or psychology of visual appeal. The parameters of solution are shock, elasticity, strength, torque, etc. This is the type of item that the artist-designer rarely gets involved in.

The typical product design class, however, is oriented toward consumer items: things designed and made to please people and to better the environment in particularly the visual and ecological areas. Thus, the artist-designer must be particularly sensitive to the basic art elements, such as color, form, size, shape, etc. And sensitive also to the effect his product has on the total environment and the psychology of the potential buyer. The student will benefit greatly if he will submit his original concept to exacting scrutiny. He must be objective enough to keep his "cool" under the most exasperating (often ridiculous) criticisms from fellow students. Remember, it is easy to see all the good things about your idea and disregard all criticism that hurts, just because it's "your baby." Students are usually more outspoken and harsh than teachers when it comes to evaluating projects. Most instructors know that a group of students grading each other's presentations will grade much lower than the teacher. Instructors are usually more tolerant because of their maturity and can often see potential in presentations that is entirely missed by the inexperienced.

The majority of students do not mince words, and the more aggressive ones often poke fun at or antagonize the student making the presentation with remarks that bring laughs from the class. Instructors usually stop this kind of behavior and point out to the heckler that this type of criticism is non-constructive, impolite and rarely offers a solution for anything.

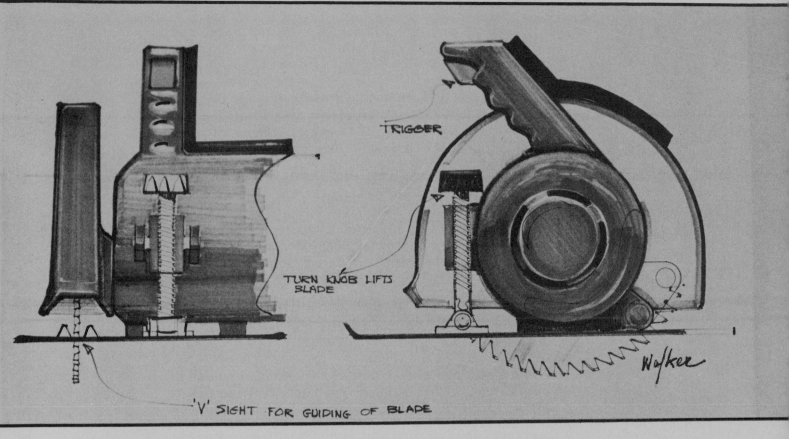

TRIGGER

TURN KNOB LIFTS BLADE

'V' SIGHT FOR GUIDING OF BLADE

Walker

Typical set of questions that might be asked during an early crit.*

1. Would a retractable guard on lower part of blade be necessary ?
2. The angle of the handle looks as if you pushed the saw forward you would tend to rock the back of the base upward.
3. What diameter is saw blade ? What H.P. motor ? Is it strong enough to turn a blade that big in half-inch wood ?
4. I like the wide base shoe. I have a saw at home which has a base only about two inches wide, and when I reach over a wide cut, like 4x8 plywood, it flexes, tilts sideways and binds.
5. Do you think a second handle would help stability on such a heavy-duty saw ?
6. Will electric cord have a flexible steel coil around it where it joins saw to prevent kinking ?
7. How do you change blades ?
8. I notice handle is not over blade. Wouldn't that cause twisting as it was pushed forward ?
9. How do you oil it ? What color is it ? I think dark colors don't show grease and oil smears like the light colors do.
10. I think a few more quick sketches would have made for a more complete understanding at this stage, instead of only two.

* Thanks to student Doug Walker of Univ. of Calif. at Long Beach, Industrial Design Section, for letting us use two sketches of his.

Nevertheless, it is good training for the student designer to take all of this in stride, formulate answers, and NOT fire back with an emotional tirade. Similar situations sometimes exist in the business world where ill-mannered clients or assistants give the impression that they know it all. If you can keep calm in the classroom, you can probably keep calm in the business world. This is not to imply you must give in to this type of person. It is certainly your prerogative NOT to work for people of that ilk. But getting sore and blowing your lid solves very little in the design world.

Once a student has withstood the barrage of questions and doubts thrown at his concept he will be that much better prepared to answer the same kind of criticism during the next presentation. Students must also learn to say, "I hadn't thought of that " and give the critic his due, than keep up the bluff that, "Oh, I knew that all along but didn't say so."

There is an interesting factor here. In observing students during a crit, it often happens that the person making the criticism, if answered politely, is the first to offer a solution to his own criticism. People in general like to help other people, and the spirit of a class is often the key to the excitement and enthusiasm of learning. Students who realize their criticisms are going to be received in the spirit in which they were given (to help) will very likely work like mad to insure the success of ALL projects, not just their own. Call it what you will: empathy, pride, enthusiasm, teamwork, espirit de corps ... it helps learning flower. It is always found in successful art agencies and design groups. It can happen in the classroom also if students will accept criticisms, not as a crack at their intelligence, but as a road to a valid "definition of concept."

Let us assume then that your design sketches have been presented to the group once or twice. You have sifted the criticisms and suggestions, talked over the problem with the instructor and have finally come up with a clear CONCEPT of exactly what you want your product to accomplish. No product can be all things to all people, so you must know what your design objective is within definite limitations, whether they be function, appearance, or price. A precise, written statement often helps communication here between student and group, or student and instructor. Once you have your objective established we can go on to the approach (how to produce it) which includes materials as well as production processes. But first, in Section 3, we come upon the elements that designers have to be aware of and sensitive to . . . the basic design factors.

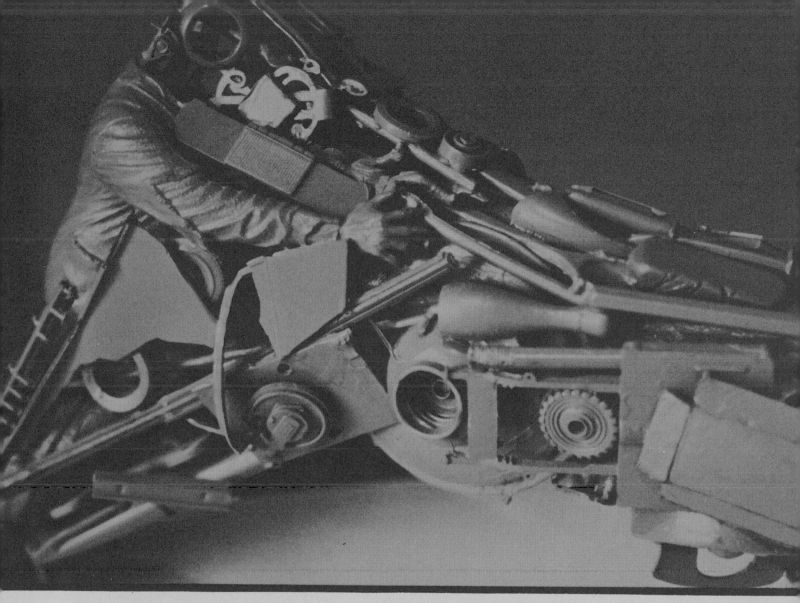

DESIGN FACTORS 3

SECTION 3   PRODUCT DESIGN FACTORS

If we design with the idea that products and machines are to be kept subordinate to human beings, then we must concern ourselves with noise, quiet color, simplicity of form, safety, and ecology as well as function, market, and price.

Design elements are those basic factors that must be taken into consideration when designing and producing a product. A number of these are outlined below and then illustrated and explored individually as basic design problems. Some of the completed 3-D units could be photographed and placed in the young student's portfolio as well as the flat design units.

Altho these units are basic, they often show a prospective employer, scholarship committee, or the school placement counselor just what you can do with color, shape, form and other relationships. At least hold on to a few for your portfolio until you get better ones. More complex and sophisticated presentations will gradually take their place. But always remember you will be hired or advanced not on what you SAY you can do but upon what you SHOW you can do. Don't throw these early problems away too soon. Start building your portfolio now.

Following is an outline of various design factors:
(This outline is also a good CHECKLIST to follow as you design.)

1.　　Visual Elements

　　　a.　　Line
　　　　　　Includes edges, axial relationships, and implied lines, such as imaginary extension of edges or center lines, as well as the drawn or painted line.

　　　b.　　Shape
　　　　　　Used in this text to describe either flat 2-dimensional forms like elevations and silhouettes, or 3-dimensional forms like cubes and spheres. It can define solids or the spaces between the solids. The solid images are usually referred to as "positive" and the spaces are referred to as the "negative" areas.

　　　　　　The outline or silhouette shape of an object or space can often be described as a geometric shape (triangle, pyramid; circle, sphere; square, cube) or as a free-form shape.

　　　c.　　Texture
　　　　　　Appears rough or smooth.

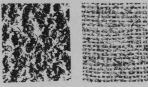

d. Value
The contrast or range of light, gray and dark.

e. Color
Hue, chroma (intensity or brilliance) and value.

2. Tactile Elements

a. Touch
Does it feel good texturally?

b. Temperature
Is heat transfer normal or irritating?

c. Grip
Does surface contour lend itself to ease of grasping or handling?

d. Resiliency
Does it give to weight or pressure? Is it too soft, too hard?

3. Kinesthetic Elements (Muscular Response)

a. Balance and recovery
How a fishing rod feels, how a coffee pot pours, or how a rocker rocks.

b. Play or resistance
The feel, for example, of a steering wheel mechanism or the difference between a "click" switch and a "mercury" switch. Cam versus flip-flop, etc.

4. Aural Elements

a. Sound
Is it desirable? Is it necessary?

b. Noise
Is it acceptable? Is it irritating?

5. Smell and Taste

Some materials give off odors, or have a peculiar taste. Is the smell good, acceptable, or irritating? Cigar boxes, cedar chests, carpeting, paneling, upholstery, wallpaper, wooden spoons, doghouses, plastic seat covers, and pillow cases are examples where odors often become part of the design study.

25

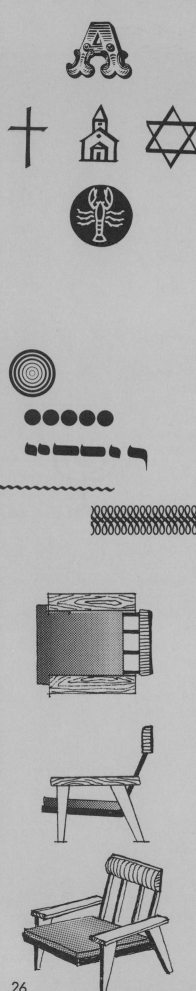

6. Nostalgic or Religious Elements

Traditional patterns, familiar images from the past, inflexible symbols must certainly be considered when designing for certain age groups. Altars, stained glass murals, chalices, courthouses, library interiors, children's furniture and even automobile designs are affected by past history and symbolism. Symbolism can be a crude, overworked cliche' as well as a sensitive piece of quick communication.

7. Utility Factors

Does the product do what it is supposed to do? Is it safe to use? Can it be serviced economically? Are spare parts available? Is it easy to store?

8. Recycling

Will the product last a long time, or is it a wasteful, "no-return," throwaway item? When it wears out can the material or parts be reclaimed? Is it necessary to reclaim it? Can it be reused in another capacity? Or is it biodegradable?

Rhythm is a concept sometimes spoken of by designers. It can probably best be described as the repetition of a design element in various ways. It is usually concerned with the visual elements. The repeat of a color here and there, of line or axial relationships, or similar textures, for example, within a design tend to give it unity or flow.

Plan view (looking straight down at an object) is a view that should be considered.

Elevation (looking straight at the side of an object) is important in silhouette as well as the shape of solid areas and spaces (positive and negative).

Three-quarter view (seeing 3 sides of the object) is another important view to consider, when judging basic esthetic elements.

In other words, an object should be designed like a piece of sculpture. It should be turned around and around in your mind as you design so that it becomes as esthetically pleasing as possible from all sides.

BASIC DESIGN PROBLEMS

The student designer often tends to think of design only in symmetrical terms. A handle is placed in the visual center of the top of a case instead of considering weight load or center of gravity. Bright colors are often used in equal areas which sometimes cause unpleasant visual "vibration." A control is positioned in the center of an empty panel even though it would look better and function better if it were placed off to one side nearer its metering readout. A sink is placed "on center" with equal drainboards on both sides, a window centered exactly over the sink faces due east, cupboards over each drainboard are of equal size and have equal shelf spacing. The early morning sun then hits milady right in the face when she is at the sink; small items don't seem to warrant using a large space shelf; large items won't fit on any shelf; etc.   The whole feeling in too many amateur design areas is the monotony of "blah" Renaissance symmetry.

People today are becoming more and more aware of "interesting" environments. Split-level homes, asymmetrical couches, contrasting colors on walls in the same room, rooms that are not at 90° with each other are examples of asymmetrical design that provide a certain "surprise." Interest is sustained through variation in size and shape of both positive (solid) and negative (space) forms. Entrance doors placed off center with pierced concrete block, wrought iron, or glassed areas providing textural variation within a pleasing Mondrian look are examples of nonsymmetrical design. Large businesses have discovered that employees work better if their desks are arranged in smaller groups instead of long evenly spaced lines of military precision. The symmetrical 9 candle candelabras are being redesigned into candleholders with arms of varying lengths and heights as well as sockets for different size candles. Even wheel-thrown pottery is being pushed and flattened into unsymmetrical forms. The traditional symmetry of design which came to us from the Greek temple, the Roman aqueduct, the Gothic church and the Renaissance civic building is no longer a sacred cow. But young designers are often not aware of this, and because they are human (two eyes, two ears, two hands, two feet) a symmetrical structure themselves, they keep plopping things in the middle just because they have never been made aware of possibilities in design other than symmetry. (Let the beginner understand here that there is nothing "bad" in symmetry. It is merely that the young designer must be aware of ALL factors of design. He can then make a design judgment based on a broad background of knowledge. Otherwise his decisions will continue to be made from ignorance, and his symmetry becomes nothing more than a nonthinking cliche'.)

This same designer does not yet understand how implied line relationships can help unify a design; how variation in size and shape can create interest; how changes in texture can please the eye or the touch; how the use of subdued chromas in color can bring relief to the eye and mind of the consumer; or how the spaces (negative areas) around or between objects are as much a part of the design as the solid (positive areas) forms are.

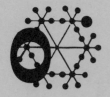

The following units then are planned to make the student aware of a few basic design factors other than "centering." The problems will also sharpen their manual dexterity IF CRAFTSMANSHIP STANDARDS ARE KEPT HIGH. Rough sawtooth edges, grubby thumbprints, exuding glue peppered with eraser crumbs, dog-eared corners, boxes sagging at 80 degrees when they were intended to be at 90 degrees are just not acceptable. The sooner students understand this, the sooner they realize a sloppy effort must be done over, the sooner the projects will become professionally presented, and the sooner the meaning of "design integrity" will be understood.

BASIC DESIGN PROBLEMS

Project 1   (Line)

On a 6 x 12 inch piece of illustration board or bristol board or scrap mat board indicate with pen or pencil a variety of closure seams used to join or edge different materials. Use a T-square to keep them parallel with one side of the board, and adjust the spaces between each line so they do not appear jammed. The entire presentation should appear neat and professional. These are often the very lines or edges that have to be indicated as part of the design presentation. A few examples are indicated below:

LACING ZIPPER HEMSTITCH WELD BOLTS BRADS TILE BASEBALL PIANO HINGE SCREWS STAPLES GIVE UP ?

Project 2 (Implied Line)

In this century of electronic controls, communication systems, and information retrieval, control units or "consoles" are often built up from standard chassis structures that fit in standard racks with standard panels or "faces." The designer should be acquainted with these kinds of components just as surely as the engineer. Whether it is the dashboard of a car, the controls of a washer-dryer combo, a radio or stereo cabinet or the walls of a space capsule — the principles of esthetic and functional arrangement are applicable to all.

On a horizontal panel, 23 inches long by 9 inches high, arrange the controls and readouts listed below in an implied line relationship. That is, the edges or corners (or sometimes center lines) of each item should be related to other items by the imaginery extension of an edge. Circular or curved items may be related by center lines or other axial relationships. Dotted lines in the example below illustrate how order can be achieved when a variety of different objects have to be combined on one surface. No item can be closer than one inch to the edge. (Why?)

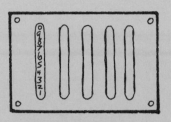

One toggle switch	1/2 inch diameter with the word "ON" above it and the word "OFF" below it in Gothic caps.
One readout panel	Horizontal 6 x 4 inches with 5 vertical number columns, one to nine plus zero, about 1/2 inch wide each
One oscilloscope face	4 inches square with circular glass tube face
Two ammeter faces	Horizontal 2 3/4 by 2 1/2 inches
Two dials	Must relate to ammeter faces, 1 1/2 in. dia.
Two dials	Must relate to readout panel, one in. dia.
Two output jacks	One labelled "Earphones". The other labelled "Recording". Each 1/2 in. dia.
One green light	Each 1/2 in. dia. Must relate to "Oscillator Input Control"
One dial	One and a half in. dia., labelled "Oscillator Input Control"
One input plug	One in. dia.
One nameplate	Horizontal, 2 in. by 1/2 in.

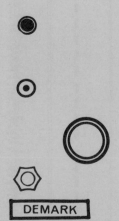

The presentation can vary from a very quick pencil sketch done on
white bond paper to a more comprehensive study done by pasting
down cut out renderings on a piece of light gray mat board which
would simulate the real enamelled instrument panel as installed in a
computer rack. Whether this problem is to be merely a "Quick &
Dirty" sketch to see if the student understands implied line relationships or whether it is meant to teach other aspects, such as delinea-
tion of knobs, dials, toggle switches, readouts, plus simulation of
colored lights, phone jacks, and plugs on a steel-gray enamelled
panel, would be up to the instructor's definition of the objectives
before the unit was started. More and more products are being
developed with control-panel surfaces, particuarly in the communication and information-retrieval areas.

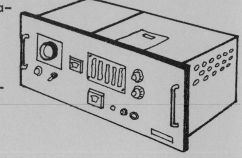

Two possibilities are shown below:

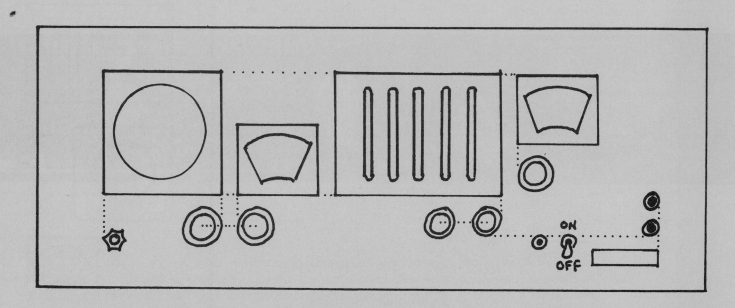

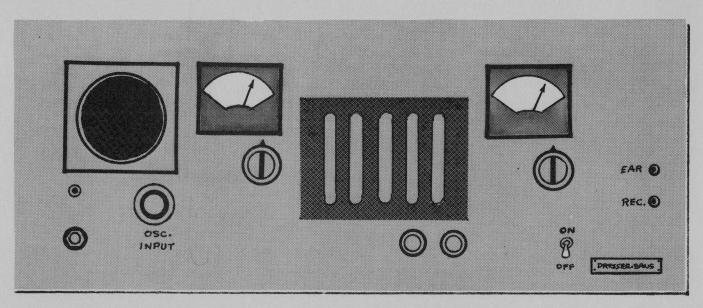

NOTE:

Obviously the placement of these controls is determined in many cases by the location of the components INSIDE the chassis. A good design job can usually be done only if the designer is in on the work from the very start and is well acquainted with the arrangement of the guts of the unit. However, the above Implied-Line problem is nonetheless a valid one in that it teaches each student how implied-line relationships help bring order to a panel that tends too often to become a jumble. Without understanding a few of these basic design factors NOW, the student can know all there is to know about the insides of the unit, all about the electronic circuits, all about the chassis and still botch the front panel arrangement, because he never had implied-line relationships explained to him.

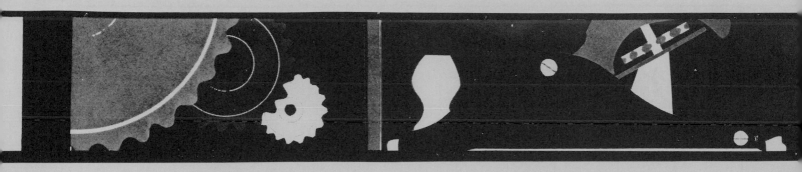

Project 3   (Flat Shapes)

On a 10 X 14 piece of subdued color mat board draw an 8 X 12 rectangle, leaving a one-inch margin on all sides. Divide the 8 X 12 area into 7, 8, or 9 different shaped rectilinear spaces so that each is a different size.

Select an interesting tool like a pipe wrench, a plane, or a bit & brace and make a number of different drawings from it on bond or tracing paper. Try different views or elevations, expand a small detail, or exaggerate a portion of the tool. Now redraw and change the dimensional (3-D) aspects to a flat shape. Transfer these flat images to black, brown, gray or white paper using your graphite transfer sheet or white-chalk transfer sheet. Cut them out and paste them down in the various rectilinear spaces on the board so that they make an interesting composition. Try for variation in your background (negative) areas in the same way you tried to make different shapes of your positive image cutouts.

This problem will help accentuate your awareness of how flat shapes, elevations, or outline silhouettes become important aspects of design whether they are incorporated into machines, furniture, or architecture. Below are two different answers.

A .22 caliber target rifle (1413 Super Match 54)

Positive and negative areas relate visually by implied line as well as kinesthetically. It looks good and it feels good. And it's used on a target not on animals.

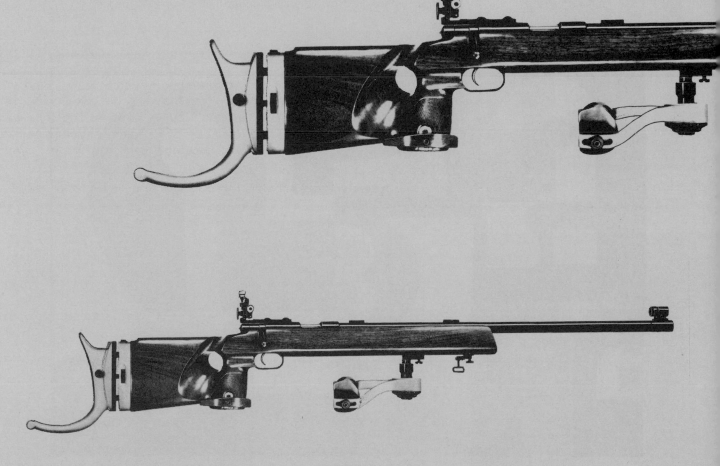

Photo courtesy Savage Arms, Westfield, Mass.

Project 4   (3-D Shape–Paper)

Create an asymmetrical architectual sculpture about 14 to 18 inches high and narrower than it is high by using a variety of cubes and other rectilinear volumes made from either white box board, bristol board, railroad board, 3-ply strathmore, or any other type of stiff snappy board that will score easily and bend without cracking or sagging. See illustration below.

Limitations:

Use between 5 to 10 boxes.
No box may be longer than 12 inches.
No box thicker than 4 inches.
Only 3 boxes may touch the ground.
Boxes must be kept clean and free from excess glue.

Suggestions:

Build a 1/2- or 1/4-size model out of white construction paper first, using a scissors and transparent scotch tape. This model can be very crude. It is a "quick-&-dirty" three-dimensional sketch, just for your own quick look. See illustration below. This can serve as a guide while you construct the real mockup or "comp."

Quick-drying Duco model airplane cement is one cement that holds the boxes together securely. There are other contact cements which can be spread on each surface, allowed to become tacky, and will hold securely when pressed tightly together. Any slow-drying glue, like white glue, is almost impossible to use, because the tabs and sides of the boxes tend to pull apart before it sets. Rubber cement is unsatisfactory, also, because it does not resist tension.

Leave extremely large tabs on the boxes or they will tend to pull apart when glued.

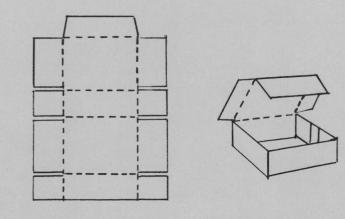

34

When scoring the corner edges of the box, lay the fold line of the tab or corner over an 1/8 - inch slot (made by laying two pieces of illustration board side by side, 1/8-inch apart). See jig illustration below. Then hold a thin strip of metal about 1/16 - inch thick over the fold line on the box and pound it down with sharp blows from a hammer. This makes a neat crease on the fold line, and the side can be bent inward at 90 degrees without cracking the box corner edge. These thin strips of metal can usually be found at a local welding shop.

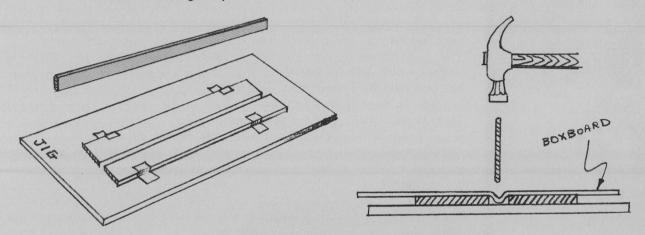

Evaluation:

1. Are the positive images (boxes) varied in size? From all views?

2. Are the negative areas (spaces between the boxes) varied also? As viewed from all sides?

3. Is the silhouette outline of the whole structure interesting from all four side views? Or is it sort of a mushy oval with no interesting balconies, bays, see-throughs, and jutting chimneys?

Cut out a number of architectual motifs from magazine scrap. Doors, windows, archways, glass, railings, stairs, planters or any other interesting textures or shapes that might lend themselves to a way-out split-level apartment house would be useable. Glue these on the sides of the boxes to simulate an apartment house or home in such a way that they interlock or make a path of texture and pattern that weaves in and out of the boxes. If you are sensitive to relationships and implied lines between doors, windows, balconies, etc., then there is less chance that the building accents will look random or "spotty." Keep looking at your structure from all sides as you apply the cutouts. The cutouts do not necessarily have to be complete units or whole objects - sometimes merely a part of the structure. Think in abstract terms now. The time will come soon enough when your design will be limited by real functional doors, and whole windows, and staircases that always start at the bottom. But unless you can appreciate the shapes and values of abstract design in your basic work, you certainly will not be able to use those talents when your designs start being compromised or limited by functional requirements.

This fountain-sculpture in San Francisco's Embarcardero Plaza was designed by Montreal's Armand Vaillancourt.

Photo courtesy of San Francisco Convention & Visitors Bureau

Photographer: Kathy Neidert

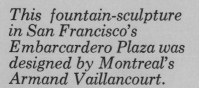

N. JENSEN

Project 5   (3-D Shape-Wood)

Find a picture of or make a drawing
of a piece of complex transportation such as:

Old-Fashioned Steam Locomotive
Helicopter
Spanish Galleon
Sailing Sloop or Ketch
Old Automobile

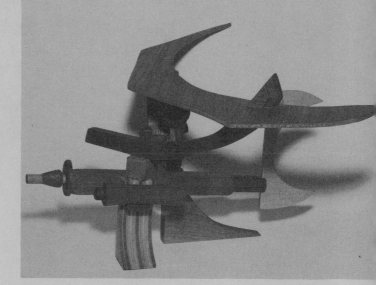

Orbohm

Create a "toy" sculpture with rectilinear or cubical wood pieces
glued together ASYMMETRICALLY as viewed from all sides. Accentuate the horizontal and vertical and keep diagonals at a minimum.
Diagonal accents may be used where necessary for items such as
sails, connecting rods, guy wires, etc., but in general, keep diagonals at a minimum. Human beings are a "square" species. We
tend to live in square homes, square rooms; sit in square chairs; cook
on square stoves; wash in square sinks; drive symmetrical, square cars.
Curves and diagonals are used occasionally for blending contours,
but on the whole too many diagonals make the average person
uneasy. Diagonals imply instability.

Suggestions:

Bring to class a variety of woods. Scraps of pine, fir, ash, birch,
and oak will give a variety of grain in the light woods. Scraps of
redwood, cedar, mahogany, and walnut will give you some darker
values for contrast in your "toy".

A bandsaw and belt sander speed things up a bit. But a good vise
with crosscut hand saw, fine sandpaper , and plenty of elbow grease
will do the trick too. A metal hacksaw gives a cleaner cut on small
pieces of wood.

Small "C" clamps will help hold blocks together if white glue is used.
The quick-drying or extra-quick-drying model airplane cement can
still be used here also.

Finish the blocks as smooth as possible
with extra-fine sandpaper or
garnet cloth before gluing together.

Delman

Arumbulo   Casner

When finished apply one or two coats of pure linseed oil.
This brings out the grain and natural warm values in the various
woods. Varnish tends to give it a glossy, harsh look. One caution
here: Be sure to let the glue dry for 2 or 3 days before applying the
linseed oil, as the oil tends to loosen the glue joints if they aren't
perfectly dry. Don't saturate the joints with oil finishes, as the oil
can cause separation, even after glue has set.

Keep viewing your project from ALL sides as you build. Remember
you are trying to AVOID symmetry. Your blocks (positive images)
should be varied in size and placed off center, and the spaces between the blocks (negative areas) should also be varied in size and
spaced asymmetrically. Wheels do NOT have to be opposite each
other. Neither do they have to be same size, and they can be square.
Think asymmetry FIRST, function second.

Evaluation:

1.   Are your blocks varied in size? In shape?

2.   Do you have some light and some dark woods? Are they spaced
      at random or are they related by implied line or a value path
      weaving through the sculpture?

3.   Are the SPACES between the blocks of wood varied in size?
      In shape?

4.   Is the SILHOUETTE outline of the whole sculpture interesting
      from all four sides? Or is it a simple rectangle with no bays
      or chimneys?

Taylor

Mesloh

Project 6   (Support Structure)

SUGGESTED MIDTERM EXAM PROJECT

The following is a short project (6 to 9 hours). The instructor will answer technical questions regarding fabrication and assembly but will not give any assistance regarding esthetic considerations. Quality of wood, size of parts, variation in negative area spacing and finish are all examples of considerations that must be answered by the student from his experiences in the previous projects.

The problem is valid in that it seems to be a fair test of whether the student can follow specifications; can put into practice his knowledge of positive and negative spacing; can solve a simple engineering problem of rigidity and stability regarding an irregular object and also meet a definite deadline.

Each student selects an interesting rock which is at least 6 inches in thickness or breadth and probably not longer than 12 to 14 inches. This should be checked out with the instructor BEFORE starting the midterm. This does not mean bringing in a rock five minutes before the midterm starts, because if it is not satisfactory, the student has no chance to find another. No professional designer would wait until the last moment and then bring in only one possibility. As a designer you must learn early in the game that you do not satisfy only yourself. You must allow the client (the instructor in this case) certain variations in selection. These selections, however, can certainly be within your proposed range of solutions.

Specifications:

1. Build a wood base for the selected rock which will support it at least 3 inches above the ground.

Willard

2. The base may have only three "legs" or contact points with the ground to prevent wobble; if a base is used it must be stable.

3. The upper part of the base must grip or partially encompass the rock. In other words, don't just build a table for it.

4. The wood base must be smooth and have a finished "craftsmanshiplike" look with neat, steady joints. Its smooth finish should be in contrast with the coarser rock texture. (DO NOT USE BRANCHES OR OTHER "NATURAL" WOOD FOR SUPPORT STRUCTURE.)

Evaluation:

1. How close have you adhered to the specifications? If you have been released from one of the specs or allowed a modification, you must get a written note ("Change order") from the instructor which must be presented on critique day to support any criticism from other students.

2. Is the base wobbly or firm? Does it grip the rock firmly?

3. Does the base look strong enough to support the item, or does it look weak and undersized?

4. Does the base look too large? Does the base become more important than the object supported, or does it enhance the object?

5. Are the positive and negative areas interestingly arranged? From all sides?

Danley

41

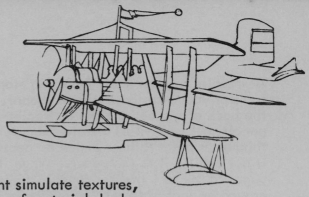

Project 7   (Texture, Value & Color)

This project's objective is to have the student simulate textures, values, and colors by painting from a variety of materials he has selected.   Then later, in more sophisticated presentations, he will be able to imitate any surface he wishes with paint, pens, or markers.

Step One:

Do a bit of research and look up some photographs or drawings of a few pieces of complex machinery. Some examples are listed below:

Triplane
Clock works
Old-fashioned automobile
Ditch-digging machine
Printing press
Steam shovel
Threshing machine
~~Automobile fender die stamping machine~~
~~Cross section of hydro-turbine electric generator~~
~~Catalytic cracker installation~~

(These last 3 are probably too complicated. And when the "way-out" textures are added, the image would become chaotic.)

If you trace out of library books, place a sheet of heavy, clear acetate over the book illustration and place your tracing vellum over that while you trace. Then you will not leave grooves on the illustrations in the book. Select drawings or photographs that have interesting silhouette shapes. Also try to find machinery that has a number of gears, levers, connecting rods, rocker arms, ratchets, ladders, steam domes, whistles, universal joints or bearings to make the drawing interesting. You can enlarge or reduce certain parts, if you wish, to gain interest. Put these parts together to make an interesting silhouette shape and transfer them to a piece of subdued color matboard. Moss green, fox brown, Gibraltar gray, burnt orange, gold, yellow ochre, are all low chroma matboard colors that would be a good background for this problem.

A suggested image size is 9 X 12 centered on the 10 X 15 toned matboard, leaving a margin all around for later matting underlap. But the size will be rather dependent upon the proportion of your machinery drawing. It may be square, long, or rather high. Don't make it too large, however, or it will take too long to paint.

A graphite transfer sheet (most art stores stock them) will help you transfer your drawing to the matboard. If you do not have one, darken the back of a sheet of bond or tracing vellum with graphite (a 3H pencil will work also), and then dampen a cloth with rubber cement thinner or paint thinner and wipe gently over the graphite to smooth it out and fix it to the paper. When it is dry it makes a good "carbon" paper to use as a transfer sheet. If you are transfering on to a dark ground or dark matboard, cover the back of the transfer sheet with white chalk or white pastel.

Step Two:

Choose a variety of textures from magazine scrap that you think you might need knowledge of or use later in product renderings or presentations. Brick, enamel, stucco, stainless steel, upholstery material, cast iron, various woods, ceramics, and formica might be examples worth learning.

Try to get some dark and some light values of the same material so you can show form on the machinery by placing the dark value on the shadow side of the object and the light values on the lighted side of the object.

Step Three:

Cut out different texture samples to conform to various parts of your machinery drawing and glue them down neatly with mucilage or rubber cement right on the drawing on the matboard. There should be no attempt to make everything look like iron or steel. Place the textures where they look interesting. A flywheel may receive a brick or enamelled texture; the shovel of the ditch digger might be covered with upholstery material, etc. Let your imagination loose for a little while and come up with a sort of a fantasy machine you might find in a child's book or a "Rube Goldberg"* comic strip.

The matboard color can show through here and there for interesting negative spaces. Try to alternate values from light to dark so the structure or form of the image will appear three dimensional.

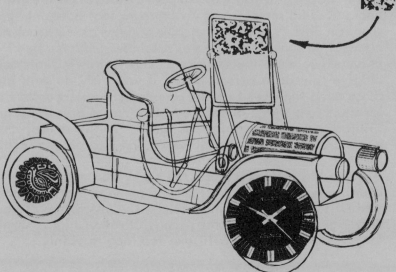

*Look up "Rube Goldberg" in library if you don't recognize the reference.

Step Four:

Trace the original drawing on the second, same size, piece of the subdued, color matboard. On this second drawing use a number 6 red sable brush (pointed) with any water-base paint, such as tempera, opaque watercolor, casein, or acrylic, and copy the texture in the pasteup as closely as possible. Use white mixed with color for light values. Do not wash in transparent washes.

Step Five:

Mat the two pieces (original pasteup and the completed rendering) side by side, about an inch apart, on another larger matboard and hand it in at deadline time.

G. WARD

45

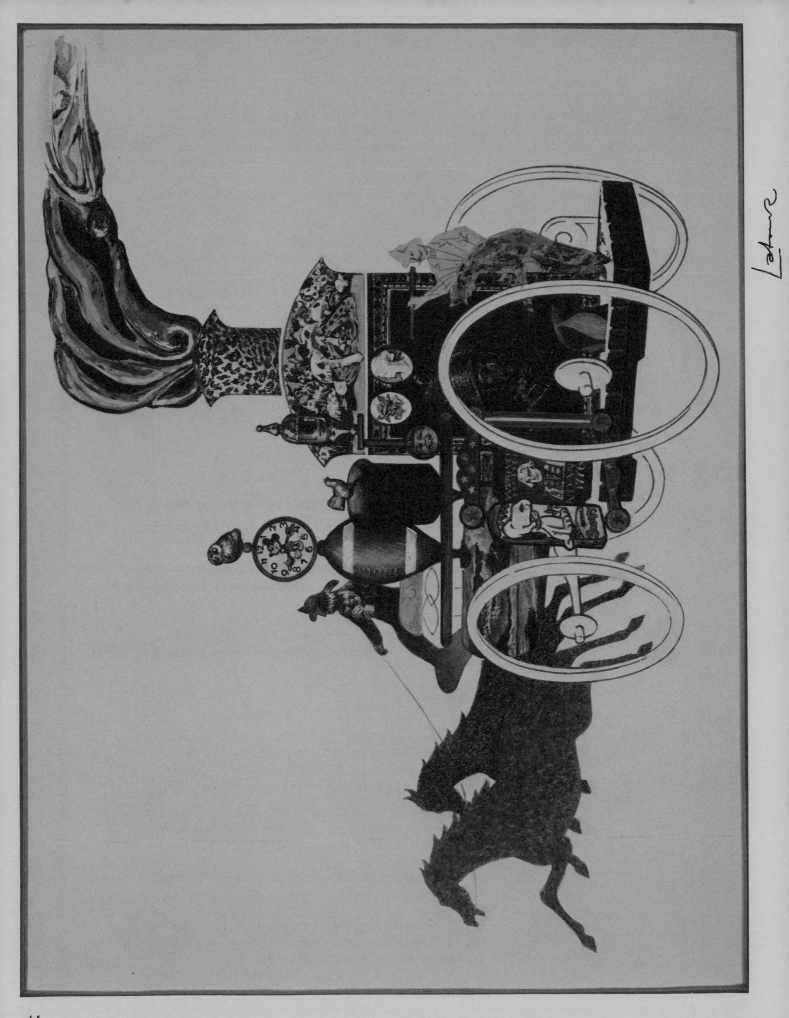

Suggestions on Color and Rendering

The chart below is a typical schematic color wheel. This one, however, has a number of color "trade" names adjacent to the block that locates it in relation to the other colors. Beginning students are often unaware of the positions of Burnt Umber, Terre Verte, Raw Umber, Venetian Red, Alizarin Crimson, or Yellow Ochre. Likewise, they may not understand the difference between Cadmium Yellow, Pale; Cadmium Yellow, Dark; etc.

One simple exercise for those students who doubt their ability to render or mix colors rapidly is to swatch in as many squares on the color wheel from the names of the colors they have, and then fill in the empty boxes by mixing the colors on each side of it. As they attempt to match colors from their paste-up textures in the previous problem they can often locate the correct HUE or CHROMA by refering to their own color wheel. VALUE can be changed easily enough by adding white to the correct HUE-CHROMA combination.

There are three factors of color to be aware of when rendering, or painting, or choosing colors for presentations. But only by actually mixing and rendering a variety of textures will you ever be able to really know CHROMA and HUE.

Movement around the wheel.

1. Hue

    The family of color. Red, Yellow and Blue are known as the primary colors and must be purchased. That is, they cannot be mixed from other pigment colors. Orange can be mixed from Red and Yellow; Green from Yellow and Blue; Violet from Blue and Red. Orange, Green and Violet are known as the secondary colors. Obviously there are many different Oranges depending on whether the artist adds more Red to it for Red-Orange or more Yellow to it for Yellow-Orange, etc. Hue is a continuous spectrum, and subtle changes in color often make the difference between an interesting color solution to a problem or a boring one.

Movement across the wheel.

2. Chroma

    The intensity or brilliance of a hue. An artist should be able to change a color from, say, bright red to a brick red or to an even duller red, like a brownish russet potato-skin color. Graying or dulling a color can be accomplished in several ways:

    a. Mixing the complement (opposite color) into the given hue will dull it. For example, mixing a little blue into orange will make a dull orange.

    b. Mixing a brown like burnt umber or raw umber into any bright hue will dull it.

    c. Mixing the 3 primaries together or 3 secondaries together will produce a brownish base. Then proceed from this base by adding any hue in small amounts.

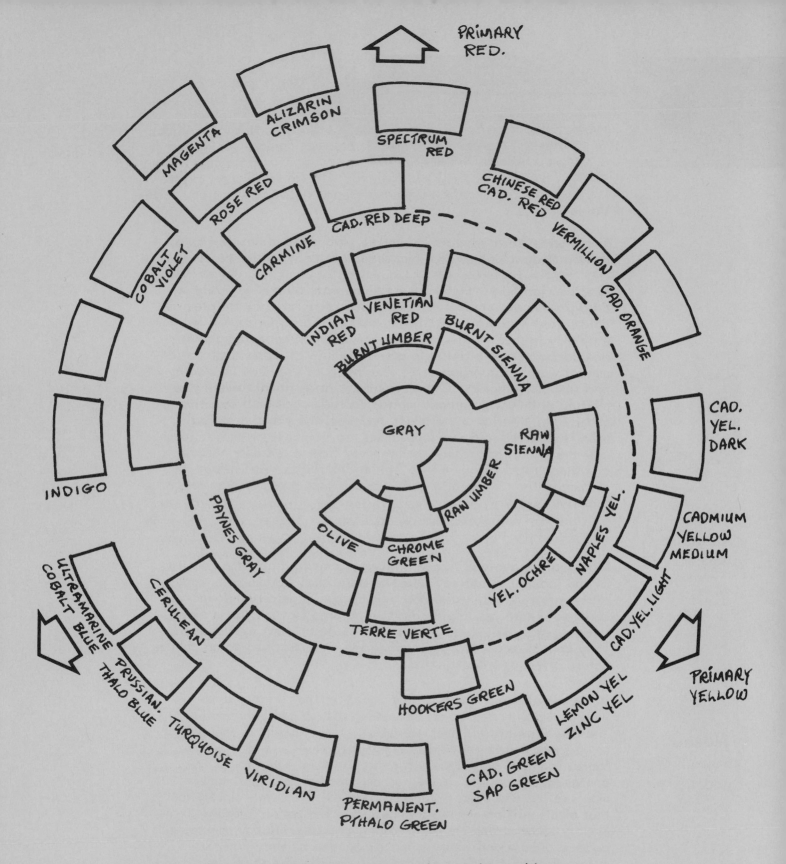

PRIMARY RED.

MAGENTA
ALIZARIN CRIMSON
SPECTRUM RED
CHINESE RED
CAD. RED
VERMILLION
CAD. ORANGE

ROSE RED
CAD. RED DEEP
CARMINE

COBALT VIOLET

INDIAN RED
VENETIAN RED
BURNT UMBER
BURNT SIENNA

CAD. YEL. DARK

GRAY
RAW SIENNA

INDIGO

RAW UMBER

PAYNES GRAY
OLIVE
CHROME GREEN
YEL. OCHRE
NAPLES YEL.

CADMIUM YELLOW MEDIUM

CERULEAN
TERRE VERTE
CAD. YEL. LIGHT

ULTRAMARINE
COBALT BLUE
THALO BLUE
PRUSSIAN.
TURQUOISE
VIRIDIAN

HOOKERS GREEN
LEMON YEL
ZINC YEL

PRIMARY YELLOW

PERMANENT.
PTHALO GREEN
CAD. GREEN
SAP GREEN

Note: Don't use black when mixing colors in this problem. Black is a "sink." That is, it absorbs light instead of reflecting it. It tends to make your colors look sooty or dead. Shadows are best achieved by using dark complementary colors. If you need a deep dark use a mixture of burnt umber plus a dark blue, or dark red, or dark green.

3.

Value

The lightness or darkness of a color. A dark blue, for example, can be made a lighter blue by gradually adding white to it.

One caution here: Light-value colors made by adding white to a bright hue are probably the most overused, cliché, tasteless colors used in industry and homes today. This unimaginative, easy way to mix a light-value color has led uneducated painters and contractors to paint pale green walls in all hospitals, mental wards, factories, and homes because they had no awareness of color chroma. One example, "institutional green" (the mixture of bright green and white), is the bane of all sensitive architects, interior and product designers the world over and is usually referred to by various names too vulgar to print. Actually the green should have been mixed first to a duller chroma and then mixed with white to lighten its value. This results in a color that is less of a garish cliché and more pleasing to sensitive people

Why?

Mankind has evolved through the ages spending his time living in and looking at nature. Most of the visual world—dirt, sand, rocks, grass, trees, water and animals are generally dull in chroma. There are bright colors in nature, flowers, some birds, sunsets, etc., but these are small in scale or short in time span. They appear as accents to enhance the more subdued colors that nature offers as a background.

It has only been in the past few centuries that bright, long-wearing paints and colors have been distributed universally. The bright paints and dyes are great in that they give the artist or designer a stronger tinting pigment to work with and a longer range of brilliance. But the potential intensity of the contemporary palette in no way negates the fact that modern man's retina and color psychology are still tied very closely to that of his million-year-old ancestors. The retina fatigues rapidly under strong brilliant colors, and the majority of people become uneasy and weary when surrounded by strong bright colors over long periods of time with no alleviation of chroma change.

Anyone who has observed a young couple buying furniture and drapes or painting rooms in a home for the first time will note that they often select colors that are too bright, too large in area, too many, or too textured for the individual rooms. They live with them a while—but as they gain experience they usually return to subdued chromas of "monks cloth" soft wood grains, brick colors, rock and the simplicity of the more "natural" colors of earth.

A lesson can be learned here by the beginning designer: Beware of bright hues! Use them with restraint. Bright red and green paint may be cheap, but workmen and housewives that have to live with walls, machines, hardware, and appliances in screaming colors are very apt to start screaming at your product too, because subconsciously the color has been a constant irritant. And any malfunction will give them a good excuse to "Throw that damn thing out—and don't buy that brand again!"

Be aware also that a small color swatch or sample of a bright hue often looks great. But that same hue covering a whole wall or an entire product case may be pretty awful to clients who are sensitive to the environment. Whether it be a factory or a home, more and more people are becoming form and color conscious. If you are insensitive and sloppy on color details, the public may well leave you way behind. Use subdued chromas, hold your brights for accents, and help the world become less of a neon jungle.

Project 8   (Oral Presentation)

Once a week perhaps, the students may wish to set aside a half-hour period during which four or five previously designated students each bring in several (at least 3) small products (not packages or photographs) which they think are good examples of visual or tactile or kinesthetic or nostalgic design. One product may, of course, exemplify all four areas. (Photographs are difficult to assess when considering tactile or kinesthetic design factors.)

Each student then goes to the front of the class group and presents his products and the basis for selection using no more than 30 seconds per product. At the end of his verbal presentation allow about 2 or 3 minutes only for class discussion and then proceed to the next student. This is one easy way to get students up in front of a group, give a quick presentation, field a few questions and sit down. They gradually discover it is kind of fun (if they are prepared) to support one's own theory or philosophy of design against criticism. Then later they can make longer, more sophisticated presentations and take a clients's criticisms with an open mind and not panic.

OR

Designated students bring in a few products that they feel are poorly designed and in front of the group suggest redesign possibilities in terms of:

1.   Better function
2.   Better safety
3.   Easier maintenance
4.   Improvement of visual appearance
5.   Recycling possibilities
6.   Simpler fabrication or assembly

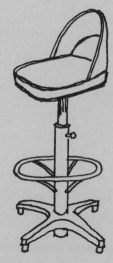

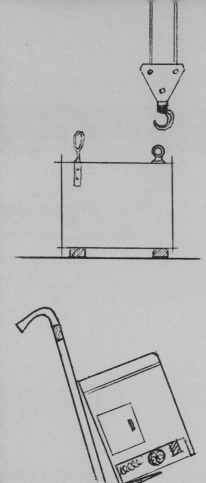

Transport Structure

When designing heavy equipment that has to be transported, don't forget to provide lifting eyes or strap loops so that cranes can lift the products onto flatbed trucks or into the holds of ships.

Too many products between the weight of 100 pounds (which is about what one man can lift without getting a hernia) and 500 pounds (which four men can lift) have no handholds, no rub rails, or fork-lift slots. Above 500 pounds the designer usually considers transportation problems, but below this weight he has been absolutely negligent.

Refrigerators, stoves, dishwashers, clothes washers and dryers usually have metal skirts which get bent all to hell if you put a hand truck under them. Absolutely no consideration has been given as to how the fans, copper tubing, grids, legs, or braces are kept from being bent when the appliance is slid over the edge of a loading ramp or up several flights of stairs. Mattresses, bedsteads, davenports and pianos are another source of "damaged-in-transit" stamps on invoices, merely because no designer has ever thought of the fact that the product has to be picked up and delivered before it is sold. And if you gathered together all the electronic technicians in the United States and asked them what suggestions they would have regarding the handling of heavy items of electronic equipment, you would probably have enough data to promote TWO design conferences at Aspen, Colorado.

Too often the designer assumes that because the thing is slid into a corrugated carton with a neat emblem on the side that his job is done. He should always consider that "Human Engineering" should concern itself with the people that transport and repair his product, as well as the people that use it.

What are rub rails? They're usually two tough, rigid bars that extend between the legs of a piece of gear, so that when the cabinet is slid over the back of a truck, or a short flight of stairs, or a concrete curb, the rub rails slide on the curb, for example, and support the base of the cabinet high enough to prevent damage to components flush with the base.

MATERIALS 4

C. WINDSOR

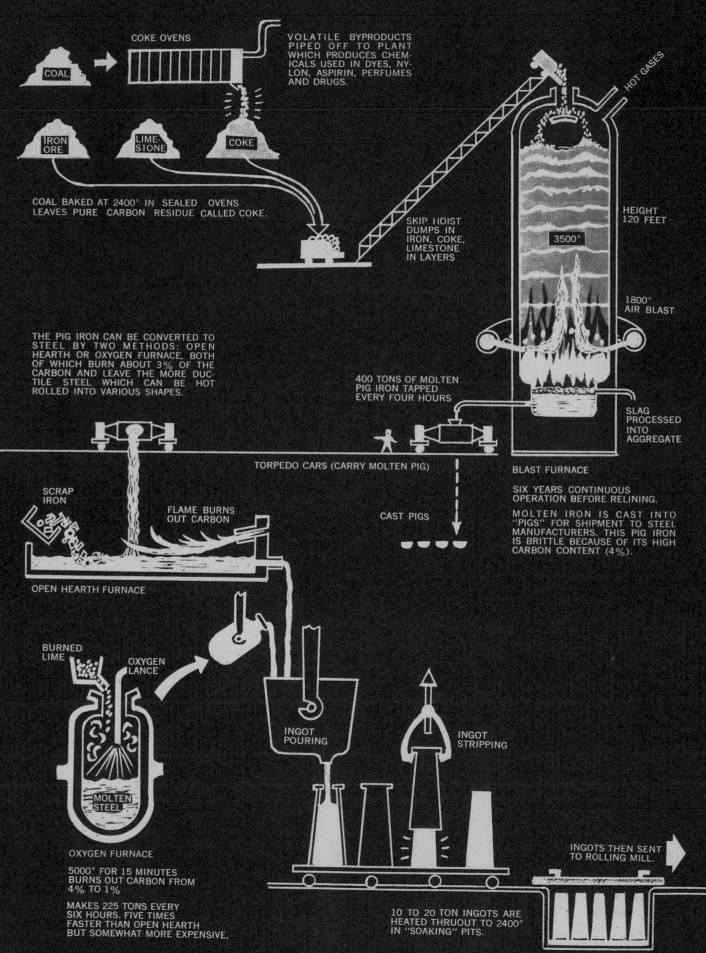

COKE OVENS

COAL

VOLATILE BYPRODUCTS PIPED OFF TO PLANT WHICH PRODUCES CHEM-ICALS USED IN DYES, NY-LON, ASPIRIN, PERFUMES AND DRUGS.

HOT GASES

IRON ORE

LIME-STONE

COKE

COAL BAKED AT 2400° IN SEALED OVENS LEAVES PURE CARBON RESIDUE CALLED COKE.

SKIP HOIST DUMPS IN IRON, COKE, LIMESTONE IN LAYERS

HEIGHT 120 FEET

3500°

1800° AIR BLAST

THE PIG IRON CAN BE CONVERTED TO STEEL BY TWO METHODS: OPEN HEARTH OR OXYGEN FURNACE, BOTH OF WHICH BURN ABOUT 3% OF THE CARBON AND LEAVE THE MORE DUC-TILE STEEL WHICH CAN BE HOT ROLLED INTO VARIOUS SHAPES.

400 TONS OF MOLTEN PIG IRON TAPPED EVERY FOUR HOURS

SLAG PROCESSED INTO AGGREGATE

TORPEDO CARS (CARRY MOLTEN PIG)

BLAST FURNACE

SIX YEARS CONTINUOUS OPERATION BEFORE RELINING.

SCRAP IRON

FLAME BURNS OUT CARBON

CAST PIGS

MOLTEN IRON IS CAST INTO "PIGS" FOR SHIPMENT TO STEEL MANUFACTURERS. THIS PIG IRON IS BRITTLE BECAUSE OF ITS HIGH CARBON CONTENT (4%).

OPEN HEARTH FURNACE

BURNED LIME

OXYGEN LANCE

INGOT POURING

INGOT STRIPPING

MOLTEN STEEL

INGOTS THEN SENT TO ROLLING MILL.

OXYGEN FURNACE

5000° FOR 15 MINUTES BURNS OUT CARBON FROM 4% TO 1%

MAKES 225 TONS EVERY SIX HOURS. FIVE TIMES FASTER THAN OPEN HEARTH BUT SOMEWHAT MORE EXPENSIVE.

10 TO 20 TON INGOTS ARE HEATED THRUOUT TO 2400° IN "SOAKING" PITS.

SECTION 4   MATERIALS

Metal Forms (Ferrous and Nonferrous)

A.   Blast-furnace product

Pigs — Crude chunks of metal made by pouring the molten metal from the blast furnace into sand molds

Ingots — Cast chunks of raw iron or steel

Billets — Rough slabs of iron or steel ready to be heated and rolled into a desired shape

B.   Shape variations made by reheating and reforming

Plates — Heavy structural slabs used for buildings, ships, armor plating, etc.

Sheets — Thinner slabs which allow for reshaping by roll forming, stamping, drawing, etc.

Strips — Sections of sheets in varying widths

Bars — Round and square stock in varying lengths

Wire — Diameters from 1 inch to thinness of human hair

Pipe — Hollow stock of varying diameters. Thin-walled pipe is known as tubing   or conduit.

Beams — Structural members of varying cross sections and strength; angle iron, C-channel, I-beams are typical examples.

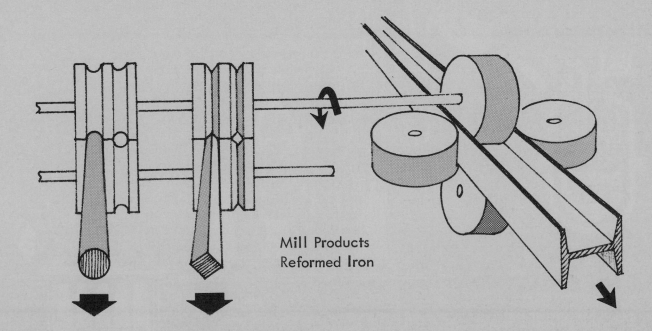

Mill Products
Reformed Iron

Metal Qualities

A.   Ferrous (Iron) Chemical Variations

Casting pigs are usually made up of 4 elements; iron, carbon, silicon, and manganese. Changing the specifications gives a wide range of characteristics. The main products are listed below:

1.   Gray iron — A 2% to 4% carbon content precipitated as graphite during solidification acts as a buffer against contraction when cooling and allows for strong unstressed sections.
   Applications: Engine blocks, machinery bases

2.   White iron — Very wear resistant
   Applications: Grinding edges, tumbling liners

3.   Ductile iron — Better elasticity. Holds up under stress.
   Application: Crankshafts, gears, hinges

4.   Malleable iron — Can be hammered into shape when heated. Flows under pressure without cracking.
   Application: Wrought-iron furniture, sculpture

5.   Carbon steels — Vary in chemical composition and have characteristics of toughness and elasticity. Probably more carbon steels are used in manufacturing than all other metals combined.

6.   Stainless — Steel that usually contains more than 11% chromium. There are many varieties; used to resist corrosion or in applications at high temperatures.

7.   Alloys — Innumerable combinations of chemicals and metals emphasize certain characteristics, such as high-temperature applications for valves, furnace parts, exhaust manifolds; or possibly corrosion resistance for use in acids, alkalies, salt water; or strength in connecting rods; elasticity in piston rings, etc.

B. Common Nonferrous Metals and Alloys

Nickel – A whitish ductile metal used in alloys and as underplating in chrome plating.

Chromium – A whitish, hard, brittle metal used in stainless alloys and plating.

1. Aluminum – Density is approximately 1/3 of steel. Resists corrosion without colored oxides. Conducts heat and electricity readily. Variety of alloys.

2. Copper – High electrical or thermal conductivity. Corrosion resistant. Weak at high temperatures. Interesting color.

3. Brass – Contains copper up to 60%, Zinc to 40%, Lead to 4%, and Tin to 5%. Shiny yellow look.

4. Bronze – Copper to 90%, Tin to 10%, Lead to 4%. Greenish brown appearance. Dull finish usually.

5. Titanium – Very abundant metal. But the need for critical production controls raises its cost. Lighter than steel, heavier and stronger than aluminum. Low thermal conductivity. Nonmagnetic.

6. Zinc – Bluish-white. Used in making galvanized iron. Resembles magnesium.

7. Magnesium – Very lightweight, ductile, silver-white. Burns with dazzling white light. Used in lightweight alloys.

8. Manganese – Hard, brittle, grayish-white. Used in steel alloys to give toughness.

9. Tin – Low-melting. Similar to silver in lustre and color. Used in plating other metals.

C. Refractory Metals

Melting points above 3600°F. Available in regular mill products and also in bolts, screws, wire, etc.

1. Tungsten – Twice as heavy as steel. A rather rare but very important metal. Gives steel the toughness for high-speed cutting tools. Lustrous gray color.

2. Tantalum – Very acid-proof.

3. Molybdenum – Silverish white. Used to make steel hard. Resists hydrofluoric acid especially.

4. Niobium – Also known as Columbium. Superconductive.

PROPERTIES OF MATERIAL

1. Malleability — How much it can be permanently deformed by compression without fracturing.

2. Ductility — How much it can be changed in shape without breaking.
(For example, lead is very malleable when beaten but tears when stretched, so it is not ductile.)

3. Hardness — Its resistance to penetration. (Testing machines drop points or ball-shaped weights on the metal and the depression made in the surface is measured against a standard scale such as Rockwell C.)

4. Toughness — The ability to withstand shock or impact without fracture.

5. Strength — Resistance to any deformation, such as compression, tension, bending, shear, torsion.

6. Elasticity — Characteristic of returning to its original size and shape after distortion. (Steel often returns to its own shape to a better degree than a rubber band, so it has better elasticity, believe it or not.)

7. Plasticity — The extent to which a material can be permanently deformed without rupture. (Soft clay might be very plastic.)

If steel can be picked up by a magnet it is called magnetic. However when a steel is heated above its critical temperature it cannot be picked up by a magnet and is called nonmagnetic. Iron can be given a magnetic field. It is then called a magnet.

The Critical Temperature of a metal is temperature that makes it become very hard and usually brittle. Charts give optimum points for all metals.

1. Hardening

Steel can be made hard by heating it to its critical temperature, letting it stay (soak) at that heat for a time, and then quenching it in liquid, such as oil or brine. The liquid may be heated to certain degrees to allow for different speeds of quenching. After cooling, the steel is very hard but is brittle and fractures easily because of internal stresses set up by the heating. Depending on soak time the hardening can be shallow or deep.

Shallow hardening is known as CASE hardening.

2.   Tempering

To relieve these stresses the metal can be reheated to a tempering level (lower than the critical temperature) usually right
after the hardening process, and then cooled more slowly.
This causes a slight decrease in hardness but increases the
toughness. Cold chisels are a good example of hard steel that
can cut other metals but tough enough not to fracture when hit
with a sledge hammer.

3.   Annealing

Hardened steel may be softened by this type of heat treatment.
Annealing appears similar to tempering but softens the metal
to a greater degree and thus improves its machinability and
relieves internal stresses set up by cold-working operations.
Usually accomplished by slow cooling after heating.

Hardness Testing

Rockwell Scale C is a range of hardness from 0 to 68. A
rough test of hardness may be done by cutting the metal
with the corner of a sharp file.

Hardness No.

20 File cuts metal easily.
30 Metal resists cutting.
40 File cuts metal with difficulty.
50 File barely cuts metal.
60 File will not cut metal.

Mohs Hardness Scale (Usually Mineralogy)

Can be scratched with:

1.	Talc	Fingernail
2.	Gypsum	Copper coin
3.	Calcite	Knife blade
4.	Fluorite	Glass
5.	Apatite	Abrasives
6.	Feldspar	Crocus (Iron Oxide)
7.	Quartz	Flint
8.	Topaz	Garnet
9.	Sapphire	Emery – black corundum (aluminum oxide plus iron oxides) similar to emerald
10.	Diamond	The Mohs scale is roughly arith-metical to step 9, but diamond, at step 10, is 40 times harder than sapphire.

Heidi

PLASTICS

Forms include those similar to metals:
plates, sheets, strips, bars, and pipe.

Thermoplastic resins may be softened, hardened and then
resoftened without changing chemical compositions.

Thermosetting resins can be formed by pressure or heat, but
this causes a chemical change in composition and the resultant
form can not then be reheated, softened and reformed.

Plastics tend to be:

a.   Heavier than some woods
b.   Not as elastic as rubber
c.   Not as transparent as glass
d.   Not as scratch resistant as steel

Yet they are:

a.   Lightweight
b.   Transluscent or clear
c.   Flexible
d.   Corrosion resistant
e.   Waterproof
f.   Nonconductive
g.   Easy to fabricate

Disadvantages in some respects are:

a.   Tendency to discolor, crack, scratch or become brittle
b.   Higher cost material than metal
c.   Low heat resistance
d.   Low strength in some applications
e.   Not biodegradable

Plastics and glass, which is a plasticlike material, can also
be formed by molding, casting, rolling or extruding. (See
Section 5 for description of these processes and others.)

There are an infinite number of "plastic" materials. In general they are made from giant molecules called polymers. They are the first man-made materials. For thousands of years man was dependent on what he could do with stone, wood and metal. Then in the early 1900's celluloid and hard rubber came into use. The yellowed celluloid windows in convertible tops were a familiar sight as were the hard-rubber, black bakelite radio panels at that time. In a sense, the alchemist's dream has started to come true. There may come a day when we can manipulate atoms as we do the large molecules and actually produce whatever material we want.

The modern chemist today can give a manufacturer plastics with almost any specified characteristics. The number of trade names labelling various plastics is staggering and very confusing to anyone trying to see the entire picture. One could argue for days regarding which is which. And your conclusions would undoubtedly vary depending on whether you were discussing the problem with a chemist, a structural foam supplier, an adhesive distributor, a paint company salesman, a synthetic rubber manufacturer, or the engineer in a fabric mill. The very word, resin, for example, which used to mean any gum exuded from trees, is now a sort of catch-all classification for a variety of plastic materials. Synthetic resins have come to mean polymers in general to the layman.

The classification below makes no claim of chemical engineering truth. It is only a list of common articles of manufacture, from hard to soft, with a current plastic chemical name and several trade names attached to each group. The student can at least then recognize that the dark-brown, hard, brittle light socket with the chain hanging out of it is probably a phenolic plastic; that the squeeze bottle his baby sister sucks out of is probably made from the same base, polyethelene, as the family's flexible garbage cans. There is a great overlap in all the classifications. So the only help the list may be is start the beginning designer thinking about and remembering from his experiences what different trade plastics will do and keep his backlog of information current. Then when he is confronted with finding a particular plastic to do a particular job he will have some start for comparison.

P = thermoPlastic
S = thermoSetting
Maybe

		Plastic	Trade Names	Uses
P	1.	Acrylic	Acrilan, Acrylite, Lucite, Plexiglas	Clear plastic, tail lights, piped-light devices
S	2.	Phenolic	Bakelite, Durite, Aqualite, Arcolite	Utensil handles, lamp sockets, electrical insulation
P	3.	Polypropylene	Excon, Olefin, Poly-Pro, Olefil	Fishing tackle boxes, rigid heat-sterilizable bottles, pipe fittings, toys

s	4.	Epoxy	Araldite, Devran, Maraglas, Fiberglas	Circuit boards, coatings, structural foams, skiis, boats
P	5.	Nylon	Capran, Chemstrand Filon, Zytel	Tennis racket strings, fishing lines, gears, drawer slides, brush bristles
s	6.	Polyester	Vibrin, Mylar, Duolite, Dacron	Tubing, boat hulls, auto bodies, panels, laminating vinyls, fibre thread
P	7.	PVC	Duran, Naugahyde, Vinylite, Boltaflex	Pipe, garden hose, phonograph records, inflatable toys, dolls, wire insulation, upholstery material
P	8.	Cellulose acetate	Celacloud, Cellon Fibestos, Kodapak	Eyeglass frames, flexible combs, packing for sleeping bags
P	9.	Polyethylene	Olathan, Ameripal, Durethen, Dylan	Squeeze bottles, milk containers, garbage cans, flexible ice cube trays, tarps, film for bags and wraps
s	10.	Polyurethane	Arothane, Carthane, Daycallan, Vulkallan	Thermal insulation, foam for mattresses, pillows, cloth backing, varnishes

Ⓒⓟ

There are other groups of plastics, such as: ABS, Acetals, Fluorocarbons, Polystyrenes, Teflons, Silicones, and Urethanes, all of which have particular characteristics that make them suitable for certain design applications. However, there are also many overlapping applications between each type. Again, the best way for the young designer to become acquainted with a variety of plastics is to visit several different plastic plants; see the machines in operation; see and handle the plastics themselves; discuss a few of their characteristics with the plant manager, if possible. Merely reading about the plastics will NOT give you the kind of information you get by actually handling the specific plastic.

Following are descriptions of five different plastics which may give the student an idea of the many parameters (factors) involved in choosing a specific plastic for a specific job.

ACETALS

Clean white color. Strong, hard, dense and crystalline with high fatigue endurance, good resiliency, toughness under repeated impact and chemical resistant. Susceptible, however, to burning, weathering, and radiation.

Use: Mechanical parts, such as cams, gears, sprockets, chain bearings, casters, leaf springs, levers, knobs, as well as cover plates or housings. Light-duty gears have an average of five times greater life expectancy than bronze gears. (Bet you wouldn't have guessed that. Eh ?)

ACRYLICS

Transparent. Cast sheet is strong and easily formed. The molding powders are adaptable to many intricate shapes. Good dimensional stability up to 250 degrees F; transmit light; are inert to most chemicals and will resist weathering discoloration even when exposed to corrosive atmospheres, salt spray, acids, alkalies, or grease. Will take metalizing, spray painting, and hot stamping. Intaglio (recessed) rear surfaces can provide illusion of a three-dimensional item buried in the plastic when it is viewed thru the transparent front surface.

Use: Outdoor signs, architectural panels, skylights, furniture, display shelves and racks, tail lights, illuminated control panels, lighting fixtures, and visual dispensing machines. Paints are made by mixing pigments in liquid acrylic.

TEFLONS (Fluorocarbons)

Antistick. Very stable at high temperatures and remain tough at low temperatures. Low friction and flexible.

Use: Linings for frying pans or cookie tins, gaskets and seals for high-temperature or corrosive service. Diaphragms which need long life flexibility. Bearings, pressure-sensitive tape, electric cable insulation. Printed circuits can be laminated in the Teflon to shield them from gas or moisture penetration but still remain flexible so they can conform to various configurations inside the rack or computer.

SILICONES

Their structure is quartzlike and almost unaffected by heat, moisture or chemicals. Relatively expensive.

Use: Varieties of thermal and electrical applications, such as fuse assemblies, coil forms, and terminal strips. Various glues and "putty" compounds for sealing in extremely high heat or icing conditions. Delicate electronic parts can be encased ("potted" or "encapsulated") in the clear liquid-type gel. This firms up into a flexible case around the components which protects the connections from heat cycling, shock or vibration, yet allows the joints to be moved or inspected visually.

URETHANES

One of the most common forms of foam plastics. Available in flexible or rigid foams. Absorb very little moisture; have a low thermal conductivity; tend to yellow when exposed to light.

Use: Flexible foams are used for upholstery cushions, clothing insulation, sponges, air filters and in packaging for protection against shock. The rigid foams are found in refrigerator panels, tank-car insulation or architectural panel structures where both mechanical strength and insulation are necessary. The denser, rubberlike urethane is one of the most abrasive-resistant materials known.

OTHER DESIGN FACTORS

1.   Fillers

Plasticizers may be added to make plastics more flexible or easier to process. Rubber added to some plastics, such as polystyrene, gives them higher impact strength. Added aluminum increases thermal conductivity and improves dissipation of static electricity. Molybdenum disulphide tends to improve wear resistance at high speeds and heavy loads. Cheap inert fillers, like asbestos, can be used sometimes to reduce cost, particularly of reinforced resins, and in some cases reduce mold shrinkage. Fire-retardant material can be added to plastics having high flammable characteristics. Graphite can be used with nylon, fluorocarbons and styrenes to decrease friction. Iron can provide radiowave shielding. Some phosphorescent pigments and other chemicals can reduce ultraviolet discoloration.

2.   Reinforcement

Reinforcement improves tensile strength, whereas fillers do not. Glass fibres are probably the most important reinforcing material for plastics. Different size filament fibres make for different characteristics. Metals in the form of wire, mesh or cloth; as well as paper, cotton, nylon or other fibrous material, like asbestos, may be used also, depending on application.

3. Corners

Sharp corners probably cause more failures in plastic parts than any other item. Design your corners with as large a radius as possible. If you do need a sharp corner try to hold the radius to a minimum of .02 inch at least. (See drawings.)

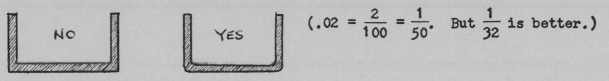

$(.02 = \frac{2}{100} = \frac{1}{50}$. But $\frac{1}{32}$ is better.)

4. Taper

Taper or "draft" facilitates easy removal of the part from the mold. Even as small an amount as 1/4 degree is sufficient. (See drawing.)

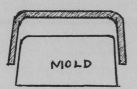

5. Holes

If you can mold through holes or blind holes (holes that don't go all the way through the material) instead of drilling them later, you can usually save money. In general, the distance between holes, or between a hole and a side wall, should be at least equal to the diameter of the hole. Provide for a minimum step of at least 1/64 of an inch at the open end of flare holes. (See drawing.) If the hole contains threads, use as coarse a thread as practicable. Threads finer than 32 per inch should probably be avoided. Don't extend your threads to the very end of the part or shoulder as, they tend to burr or strip. (See drawings.)

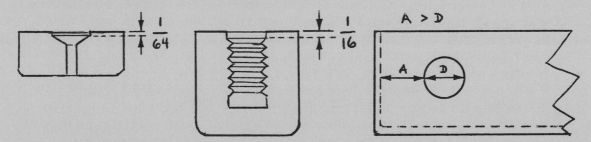

6. Expansion

Some plastics, particularly acrylics, expand and contract as much as 10 times more than glass. Therefore, any holes made to accept bolts, for example, must be dimensioned to allow plenty of "slop" or space around the bolt. Otherwise the plastic will crack when changes of temperature cause it to expand or contract.

7. Finishing

Don't forget to allow an extra thickness if any of the parts are to be machined later.

Plating (PLASTICS & METALS)

If parts of a product are to be plated for better wear, corrosion resistance or visual appearance, the designer can achieve a more economical and uniform finish by observing the following:

In general, sharp edges and protrusions steal a larger share of the current and thus get a heavier deposit. So round the edges and fillets to at least 1/64 inch, or better, 1/32 radius and avoid sharp recessed or concave surfaces. A slightly convex surface is better than a large flat surface.

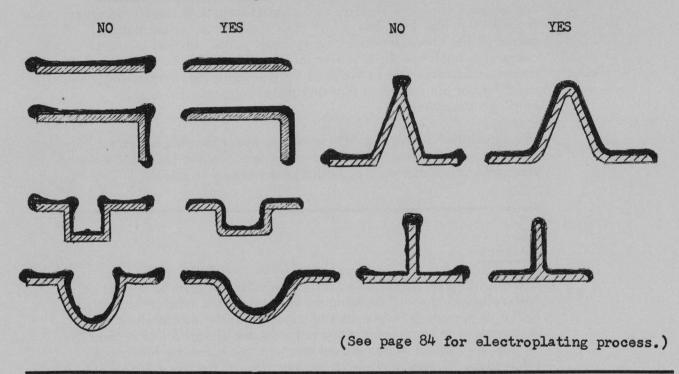

NO   YES   NO   YES

(See page 84 for electroplating process.)

STRUCTURAL FOAMS

These are generally cellular thermoplastic materials that resemble a bone structure by having a solid skin integral with an internal, light, honeycomb structure. They are made by a process similar to injection molding. Except in this case the resin is preblended with a chemical "blowing" agent which releases inert gas during the heating ejection process. The gas expands inside the mold during the shot and produces the solid skin around a rigid, cell-like porous core. The larger the molding machine, the larger the shot size. Shots up to 50 pounds of material are now feasible with probable increases in the future.

As molding techniques become more sophisticated, these light-weight, rigid foams may well take the place of most thin-walled metal castings or stampings. Appliance cases, bicycle frames, cameras, business machine housings, automobile dashboards, bodies of small recreational vehicles are all typical examples being molded at present.

Structural foams are slightly thicker than plastic solids made from equal amounts of material but have from 2 to 10 times the rigidity of the solid part. They have about twice the rigidity of aluminum and about 5 times the rigidity of an equal weight of steel. However, if wall thickness drops below .19 inch, the foam structure tends to approach the characteristics of solid wall parts. Instead of sharp corners, a small specified radius (of say .125 inch) can minimize stresses. Surfaces can be finished by woodgraining, laminating, painting, or plating, much like any metal surface. Takes self-tapping screws well.

The description above is pretty general, but it should give you a nodding acquaintance with a fast-growing technique that may lower your production cost over a similar part stamped in metal.

Concrete

Cement has good compression characteristics but must be beefed up with reinforcing rod if bending (as in a bridge) or tension is applied to it. It is made from one part of cement powder to 6 parts of sand & gravel. For backyard structures throw one shovelfull of cement in on 3 of sand and 3 of pea gravel and mix in enuf water to make a thick soupy mixture. When it sets, keep it wet for three or four days, and it will set up as if a pro laid it.

Make certain any forms you make are well made and tamped down hard around the footings. There is nothing more terrifying in the world of construction than seeing 10,000 pounds of concrete (Ask a cement company what a cubic yard of concrete weighs some day.) start oozing out of a split form or squishing out from under the footings. Believe me, there's nothing you can do then. In the Los Angeles area, for example, there is a firm that does nothing but pick up concrete from jobs where the forms have collapsed. They are on a type of emergency call schedule, because they like to get there before the concrete has set up so they can scoop it up with skip loaders. Inspection of concrete forms before pouring is extremely critical on large jobs, and it is the wise inspector who asks other experts' advice on complicated construction jobs.

PIPE

In specifying or ordering pipe, the inside diameter (I.D.) is the usual designation. One-inch pipe , for example, has an opening one inch across. However, the outside diameter (O.D.) must sometimes be specified also when the pipe is subjected to unusual pressures or use where stronger wall thickness is required.

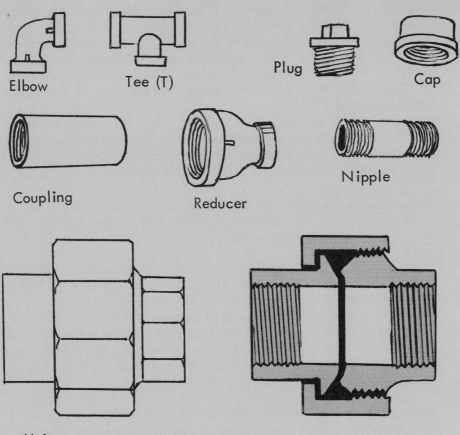

Elbow

Tee (T)

Plug

Cap

Coupling

Reducer

Nipple

Union

A pipe-joining device wherein a free-turning, threaded collar engages the threads on the mating pipe and pulls it in to seat against the face of the other pipe. Used as the last joint in a system or where two pipes must be joined without turning either pipe.

Pipe threads are usually tapered for tight fits.

Nominal Dia.

I.D.

O.D.

WOODS

Approximate order below from soft to hard.

Balsa	Porous, white, extremely lightweight.
Yucca	Fibrous, white to yellow, lightweight.
Pine	Soft, fine grain, strong, white to yellow color.
Poplar	
Tamarack	
Redwood	Coarse, long grain, slivers, oily inside. Resists rot & insects. Hard to apply smooth finish.
Cedar	Similar to redwood. Pleasant smell. Often white slashes of wood running thru the red grain.
Basswood	Similar to pine. Often used in box making.
Cottonwood	
Mahogany	Fine cross grain, compact, various shades of reddish brown or tan. Cuts easily in any direction. Does not splinter easily. Takes oil finishes beautifully.
Beech	
Butternut	
Spruce	Flexible, tough. Often used for boat spars or frames.
Balsam	
Cypress	Similar to pine in appearance. Lightweight, tough & stringy. Resists wet rot. Used in boatbuilding.
Fir	Workhorse of construction. Tough with coarse grain. Not springy. Ideal for beams, joists and plywoods.
Alder	
Elm	
Chestnut	
Hemlock	
Hickory	Tough, springy & often knotty. Early bow making.
Lemonwood	Smoother, less twisty than hickory. Also used in bows.
Ash	Bow making also. Tough elastic for strong bows or laminations with other springy woods. Used in cabinet making and cupboards or doorways.
Walnut	Fine-grained, dark-brown wood. Expensive. Used in fine furniture or household trim. Takes high polish.
Maple	Similar to walnut but lighter, reddish color.
Cherry	Similar to walnut but dark red color.
Birch	Light, yellowish color with spotted or streaked surface.
Oak	Very hard. Tough. Wide grain. Good flooring or threshholds.

Bamboo	A hollow cylindrical ~~wood~~ grass used in many house-hold utensils such as cups, forks, paddles, etc. Pails, planters, and long sticks. A hard, springy, durable wood almost impervious to water.
Teak	A dark-brown wood with many variations in grain running in color to lighter browns. Grain similar to mahogany. Takes beautiful finishes. Very durable. Used in shipbuilding.
Padouk	Tropical hardwoods used for furniture as well as jewelry. Mottled yellowish red.
Rosewood	Beautiful changing grains of purplish red.
Purpleheart	Takes high polish. Purple color.
Zebra	Definite striped grain of tan and black.
Vermillion	Brilliant orange. Vermillion dyes are often made from this wood.
Ironwood	These last are very fine-grained, hardwoods which lend themselves to carving. Extremely durable for such uses as pulleys.
Lignum Vitae	Lignum vitae sometimes has a dark, gray-green cast.
Ebony	Can be brown in color as well as black.

Lumber Sizes

Rough Sawn	Clean Sawn
1 x 1	3/4 x 3/4
1 x 2	3/4 x 1 5/8
1 x 4	3/4 x 3 1/2
1 x 6	3/4 x 5 1/2
1 x 8	3/4 x 7 1/4
1 x 10	3/4 x 9 1/4
1 x 12	3/4 x 11 1/4
2 x 2	1 5/8 x 1 5/8
2 x 4	1 1/2 x 3 1/2
2 x 6	1 1/2 x 5 1/2
2 x 8	1 1/2 x 7 1/4
2 x 10	1 1/2 x 9 1/4
2 x 12	1 1/2 x 11 1/4
4 x 4	3 1/2 x 3 1/2
4 x 6	3 1/2 x 5 1/2
4 x 8	3 1/2 x 7 1/4
4 x 10	3 1/2 x 9 1/4
4 x 12	3 1/2 x 11 1/4
6 x 6	5 1/2 x 5 1/2
6 x 8	5 1/2 x 7 1/4
6 x 10	5 1/2 x 9 1/4
6 x 12	5 1/2 x 11 1/4

Grades of Lumber

Clear	No knots
Select	Occasional knot
Number one	Few knots
Number two	Many knots
Construction grade	Lotsa knots

S2S Surfaced two sides

Wood shrinks ACROSS grain

Outside   Heartside   Outside

PAPERS

Approximate order below from thick to thin.

Chipboard	Heavy, gray, unfinished. 1/16, 1/8, 3/16 inch thick.
Cardboard	Various thicknesses. Usually finished one side.
Corrugated	⟋⟍⟋⟍⟋⟍⟋⟍⟋
Foam Core	Lightweight, foamy plastic filler sandwiched between two thin pieces of cardboard.
Matboard	Nonglare, textured, finished one side, variety colors.
Illus. Board	White surface one side. Smooth hot press or a velvet tooth called cold press. Colors available also.
Tag Board	Cream colored, tough, shiny, cheap, flexible.
Bristol Board	White, high-quality surface, flexible. Takes die cutting.
Box Board	Snappy, white, smooth surface. Scores & die cuts.
Railroad Board	Similar to Box Board. Usually in raw colors. Cheaper.
STRATHMORE	3-PLY OR 4-PLY. EXCELLENT FOR SMALL MODELS.
Watercolor Paper	Heavy weights good for some presentations and also model building.
Canson Paper	Toothy, subdued chroma colors. Ideal for wax pencil or pastel pencil renderings & other presentations.
Color Coated	Paper is coated with different hues in varying grades of value and chroma. Sample swatches are numbered for ease of ordering. Useful for gluing down on sides of models, mockups or on presentations to depict different subtle colors.
Contact Paper	Adhesive backed. Variety of printed surfaces such as walnut, ash, birch and other woods. Also some metal simulations. Useful for model presentation or mockup surfaces.
Tracing Paper	Heavier weights with vellum surface can be laid over drawings or fotos and receive felt marker or chalk or colored pencil. These renderings or presentations can then be mounted on a white backup surface.
Glass Vellum	Transparent, medium weight, acetate-like paper. Simulates glass or acetate. Good for backlighted displays, and packaging.
Acetates	Not really paper. But the frosted or clear is often used as tracing or graphic overlays on presentations. Thickness is expressed in mils (.003 or .004, .006, etc.) rather than weight of a basic ream (500 sheets) as paper thickness is measured.

Source of Materials List

Following is a list of materials that are often needed by the student product designer for models or research information. Let each student choose a few items from the list and then bring in the name, address and telephone number of the firms that handle that particular item locally. The telephone directory yellow pages offer a beginning, but there are many good sources of odd items in a local area that are little known to the average student, and this method soon gives the class a good reference list for use during the semester.

1.	Acrylic	17.	Doweling	33.	Paint
2.	Aluminum	18.	Electrical	34.	Paper
3.	Anodizing	19.	Electric Motors	35.	Plastics
4.	Army Surplus	20.	Epoxy Cement	36.	Plaster
5.	Art Supplies	21.	Foam Rubber	37.	Plexiglass (See Acrylic)
6.	Balsa Wood	22.	Foam Core	38.	Polyurethane Foam
7.	Black Chrome	23.	Formica	39.	Rare Woods (Tropical)
8.	Body Putty	24.	Fiberglass	40.	Resin
9.	Brass	25.	Gears	41.	Sheet Metal
10.	Brass Tubing	26.	Glass	42.	Stained Glass
11.	Brick	27.	Laminates	43.	Steel
12.	Cement	28.	Leather	44.	Stone
13.	Chrome Plate	29.	Lucite (See Acrylic)	45.	Tile
14.	Clay	30.	Lumber	46.	Vinyl
15.	Clock Parts	31.	Lighting	47.	Welding Rod
16.	Contact Cement	32.	Optical (Lenses)		

Sources of Information on Materials, Use and Testing

AISI American Iron & Steel Institute

API American Petroleum Institute

ASME American Society of Mechanical Engineers

ASTM American Society for Testing & Materials

SAE Society of Automotive Engineers

Sweets A system of cataloging products by manufacturer, trade name, and product. Firms' catalogs and brochures are bound into thick reference volumes and indexed alphabetically for the designers reference.

Published by: McGraw-Hill Information Systems
330 West 42nd Street
New York, NY 10036

Another source which might be of interest is "Machine Design" magazine's latest "Metal Reference Issue." This directory lists hundreds of producers under a variety of headings and codes which include most metals and processes. Address is Penton Bldg., Cleveland, Ohio 44113.

Sheetmetal Workers Bonus

Cut a thin piece of stiff tagboard or bristol board to a square exactly 8 x 8 inches (64 square inches). Now with a steel straight edge and a sharp mat knife cut the square into the four pieces as shown in the diagram below. Measure accurately.

Now rearrange the pieces into a rectangle 13 inches long and 5 inches wide as shown below. 5 x 13 equals 65 square inches. Where the devil did that extra square inch come from? So girls, if you ever get stuck cutting out a pattern and don't have enough cloth, just use the principle above and, voilá, you have it made.

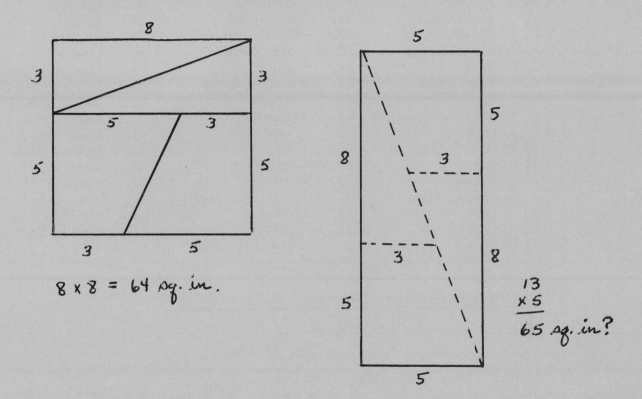

Production Processes

5

SECTION 5   PRODUCTION PROCESSES

MACHINE TOOLS

Drill Press

A drill press usually has one motor-driven spindle & chuck which can be rotated at varying speeds. The spindle can be raised or lowered by a hand lever. The table below the spindle holds the work.

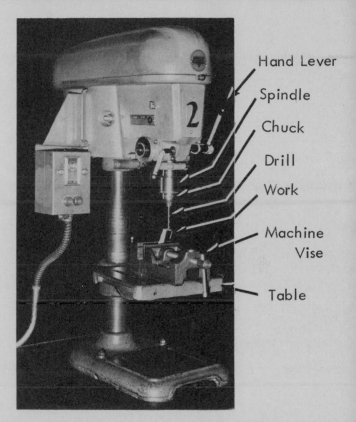

Hand Lever
Spindle
Chuck
Drill
Work
Machine Vise
Table

Lathe

Holds material and rotates it around a horizontal axis. A cutting tool is pressed against the material to shape it. The tool may be guided by hand or by mechanical methods.

Rheostat
Head Stock
Horizontal Feed

Tail-Stock Centers
Tail Stock
Tool Holder

Milling Machines

A milling machine is a very versatile tool. It is something akin to a drill press, but instead of having a simple drill spindle which drills a hole straight down thru the material, the milling machine can accommodate a variety of cutting heads on a variety of spindles. The work then is moved under the cutting heads at predetermined depths and routes to produce gear teeth, slots, grooves, insets, contoured surfaces or what ever the operator desires.

TAP

DIE

BITS

MILLING CUTTERS

REAMER

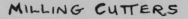

BORING

A hole is begun by drilling or trepanning and finished by reaming, boring, or broaching.

Drilling — A drill usually has two cutting surfaces at its tip. Rarely used for holes greater than one inch diameter.

Trepanning — Similar to drilling in that it cuts a thin ring of material leaving a solid core to be removed. (Brain surgeons use this method so the core of bone can be replaced.)

Extension Bit

A toothed cutting arm on the end of the bit can be extended radially up to about one inch. A hole of two inches diameter can thus be drilled.

Fly Cutter

Similar to the extension bit, the fly cutter is larger and stronger and can be extended radially for a distance of several inches for trepanning much larger holes.

In both processes above, the hole remains constant in diameter but increases in depth.

Reamer — Similar in appearance to a drill, but it has the cutting edges on the side of the tool. During reaming the hole remains constant in depth but increases slightly in diameter.

Bore — A chunk of metal with a single cutting edge which works against the side of the hole. The work is turned or revolved against the bore, which shaves off a few thousandths of an inch. By adjusting the "feed" of a bore, the machinist can produce a hole which is wider inside than at the mouth, or any other shape desired as long as the cross section is round.

Broach — Can produce oval, square, splines or other non-round holes. The broach is not rotated, as the drill is, but moves back & forth, in & out of the hole. Several broaches are usually necessary. The first broach takes off just a small amount of metal. Each successive broach has a slightly larger or "intermediate" cutting face. The last broach used has the exact shape which is desired in the hole. (By turning the work as the broach advances, the machinist can make spiral cuts, rifling in a gun barrel, etc.)

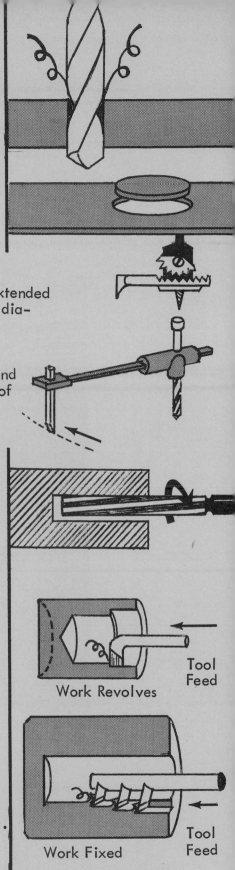

Work Revolves

Tool Feed

Work Fixed

Tool Feed

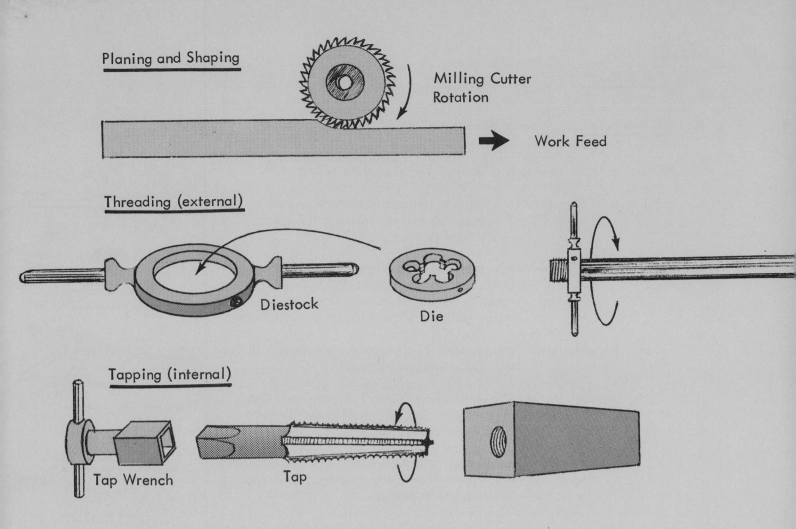

Planing and Shaping

Milling Cutter
Rotation

Work Feed

Threading (external)

Diestock

Die

Tapping (internal)

Tap Wrench

Tap

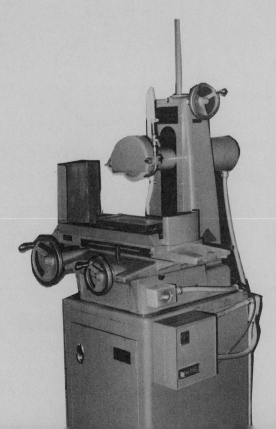

Surface Grinder

To make surfaces of steel parts smooth
and level within thousandths of an inch,
an adjustable grinding wheel makes
successive passes over the material which
is held in place on a table or base. This
table is often electromagnetic to hold
steel material tightly on level plate.

78

PROCESSES

A. **Fabrication**   A semi-manual method of forming blanks into units prior to assembly. (A blank is a piece of material cut from mill stock to accommodate the process or fit the machine.)

1. Roll Forming — Sheets or strips are bent or formed by successive rollers pressing the metal between them.

2. Stamping — Cold metal is pressed over a die form or punch, usually by a powerful hydraulic press. One stamp usually does the job. Coins, gaskets & car fenders are examples.

3. Forging — Hot material is hammered by repeated blows to produce desired shape.

4. Annealing — When metal is worked (beaten or bent repeatedly) the grains of metal tend to become distorted or broken. Stresses are set up and the metal gets harder. By low reheating (after working) the grains tend to flow in line again or recrystallize and relieve the stresses.

    a. Cooling it slowly leaves it soft and workable (malleable) again.
    b. Cooling it quickly makes it brittle and "Tempers" it.

5. Drawing — Hammering or pressing repeatedly on cold metal to form shape.

6. Deep Draw — Drawing metal out to extreme shapes like a deep bowl.

7. Spinning — Material is shaped by pressing it over a spinning die or mandrel. Saves material. Cones and reflectors are often made this way.

SEE DIVISION PAGE 74

8. Extruding — Forcing material through a nozzle or die producing a desired cross section.

9. Sintering — Heating up small particles, then compacting and fusing them into a solid. Bearings are sometimes made this way, in which graphite or oil particles are fuzed into the metal so that the bearings are self-lubricating.

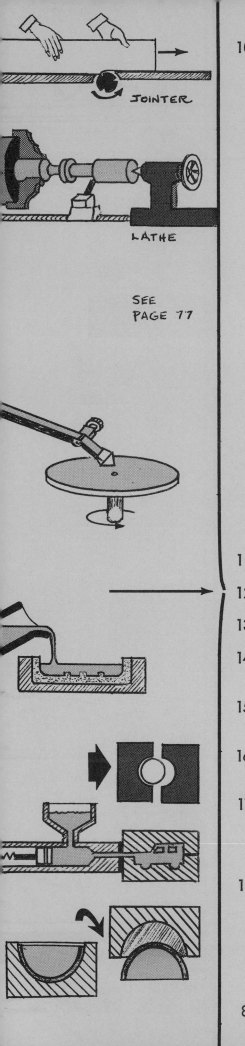

JOINTER

LATHE

SEE
PAGE 77

10. Machining — Forming parts by metal removal. Lathes, drills, jointers, or milling cutters are typical equipment.

a. Threading — Cutting external threads on bolts, pipe, etc.
Tapping — Cutting threads inside metal or nuts.

b. Drilling — Boring holes.

c. Turning — Shaping material with cutting tools while the material is rotating in a lathe.

d. Milling (Routing) — Removing material with the point of a drill device or milling cutters in certain areas. (SEE PAGE 76, 78.)

e. Reaming — Enlarging a hole by twisting a cutting tool in the opening.

f. Broaching — Enlarging a hole by forcing a cutting tool in and out of the hole.

g. Grinding — Wears the material away by friction. Usually accomplished by forcing the material against a carborundum wheel.

h. Lapping — Using an abrasive powder between two moving or rotating surfaces. Valves are "ground" by rotating the valve against the surface it seats on. Gems are lapped or faceted by holding them against a rotating disc at a precise angle.

11. Bending — Let you guess at this one.

12. Piercing — Pushing out a hole without loss of material.

13. Swaging — Hammering metal when it's cold.

14. Sand Casting — Pouring hot metal or liquid material into stiff clay-like sand molds.

15. Die Casting — Hot metal or material is forced into metal ✳ molds usually under hydraulic pressure.

16. Compression Molding — Pressing material between two dies.

17. Injection Molding — Squirting material into dies that cool or harden the material quickly and eject the finished piece in one motion.

18. Slush Molding — Pouring material into a mold, pouring it out and letting the remaining material harden as a shell which is released from the mold later.

✳ IRON AND STEEL DO NOT LEND THEMSELVES TO DIE CASTING. THE HIGH TEMP. DESTROYS THE DIE.

19. Vacuum Forming — Sucking a sheet of material over a form. Letting it set until it assumes the shape of the form.

20. Bossing — Putting a bump or knob on a piece of material.

21. Splining — Putting a ridge on a rod or other form.

22. Fluting — Cutting a slot into a rod or other form.

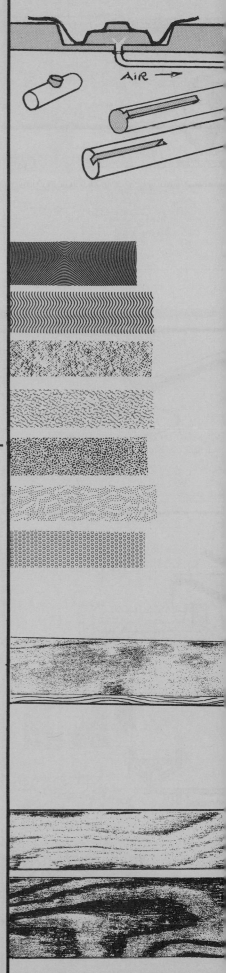

B. Finishing

1. Polishing — Buffing with soft cloth wheel or pounce often with a fine paste or fine powdered abrasive. Can become mirror-like.

2. Burnished — Rubbed with convex surface like the heel of a spoon to flatten obtrusions.

3. Brushed — Mildly scored with stiff wire brush or wire wheel.

4. Anodized — Creating oxides on surface which increase corrosion resistance and change the surface color.

5. Etched — Surface is exposed to acid fumes which leave microscopic pits for a dull matte finish.

6. Sand Blasted — Silica sand shot from an air hose roughens the surface.

7. Shot Peened — Shot hitting surface makes a pattern of small round craters all over the surface.

Finishing Woods

Filler — Often used on soft woods as a sizing to lessen its absorbing qualities and thus allow stain to be applied without its soaking in too fast or "burning" the wood black. Shellac filler should not be used on teak as it doesn't mix with the gums in the teakwood.

White shellac and alcohol (1/2 & 1/2) make a good filler for pale woods or before staining. Wax can be used over it to help protect finish and prevent staining. Shellac should not be used on Teak and probably not on Red Cedar or Ebony because the gums in these woods seem to react or "gum up" when shellac is applied in some cases.

Bleach — Used on dark or contrasty grain woods in many cases to tone down the value or contrast.

Stain — A pigment in an evaporative base. Used most often on pale or "flat-"looking woods. The stain tends to bring out the grain by being absorbed in the soft areas and repelled by the harder grain.

FLAX
SEED

BEES
ETC.

GUM
TREES
RESIN

1.   Linseed oil   Brings out the natural finish of the wood
                        and tends to make it look darker and
                        richer with a mild lustre.

Linseed should usually be cut about 50% with turp or paint thinner on first application to new wood. Let stand about 15 minutes or more between coats and then rub in linseed oil with rags or palm of hand. Six to twelve coats will give a good protective sheen. Too little will leave wood dry after it soaks down in.

2.   Wax

Waxing a surface means usually that it must be renewed every month. Wax is not as penetrating as the oils. Paste wax probably best for woods. Should not be put on over linseed oil finish.

3.   Shellac   Use alcohol as a thinner, not paint
                        thinner.

4.   Lacquer   Coat and sand, coat and sand, etc.

Lacquer is a shellac-based high-grade varnish usually applied in many thin coats with sanding in between each coat (unlike the heavier oil-based varnishes that are often applied in two or three thick coats). Used for fine interior applications such as cabinets and small boxes. Obviously must be thinned with alcohol, not paint thinner or turpentine.

5.   Varnish

Varnish is a viscid solution of resinous materials in oil or volatile liquid which dries by evaporation or chemical action. It forms a hard lustrous coating which resists the action of air and moisture. Can be thinned with turpentine or paint thinner. Often used in out-door applications such as boat decks & spars, door kick-plates, floors, and chests.

6.   Polyurethanes

Urethane varnishes and other resin coatings are akin to the regular varnishes except that they are man-made chemicals which simulate the old fashioned tree gums and resins.

The line between "finishing" a surface and "coating" it is rather nebulous. But as the finishes, such as linseed oil and thinned out varnishes, are applied more heavily and tend to obscure the original look of the material, one could call it a coating.

Before we list a few coating processes, however, it might be a good idea to put in a section on abrasives. Abrasives also fit under Machining or Grinding, as they wear the material away by friction or tiny cutting edges and are often used to smooth, polish, or "finish" a piece of metal, wood, or plastic.

ABRASIVES (Coarse to Fine)

1. Coarse Files

2. Fine Files

3. Grinding Wheels

4. Flint or Garnet Paper

The abrasives listed below range from coarse files thru the coarse sandpapers to very fine silicone carbide papers, steel wools and finally the powders.

Abrasive grades of size or coarseness and approximate equivalancies of various "sandpapers"

*Mesh size

For example, 50 mesh means there are 50 openings per lineal inch or 2500 (50 x 50) openings per square inch in the screen seive which the particles drop through. The greater the number, the finer the particles.

	Silicon* Carbide	Emery	Garnet	Flint	
Very Coarse	12		4 1/2		
	16		4		
	20		3 1/2		
	24		3	3 1/2	
	30		2 1/2	3	
Coarse	36	3	2		
	40	2 1/2	1 1/2	2	
	50	2	1	1 1/2	
Medium	60	1 1/2	0		
	80	1	1/0	1	
	100	2/0	1/2	1/2	
		0	0		
Fine	120	1/0	3/0	1/0	
	150	2/0	4/0	2/0	
	180	3/0	5/0	3/0	
Very Fine	220		6/0	4/0	
	240		7/0	5/0	
	280		8/0		
	320		9/0		
	400		10/0		
	600				

5. Steel Wool

6. Powder

Pumice   (Fine volcanic glass-like stone)
                Used with water or oil as thin paste on soft pad
                or polishing wheel. Reduces glossy look on lacquer.

Rottenstone   (Siliceous limestone)
                Less cut and more polish than pumice.

Talcum   (Mineral of a magnesium silica combination)

Tripoli   (Crushed shells of diatoms)

Rouge   (Ferric oxide)
                Very fine. Used for finishing polish on silver and such
                critical surfaces as telescope reflectors & lenses.

C. <u>Coating</u>

Can be done in a variety of ways such as brushing on, spraying, dipping or fusing (melting) on surface. The coatings can then be air dried, infrared dried, baked on, etc.

1. Paint — Pigment mixed with a vehicle such as oil or varnish.
2. Enamels — Paints which dry with a smooth, shiny look.
3. Vitreous Enamels — Glass fused to metal, etc.
4. Lacquer — A varnish usually made with a natural resin base or a combination of shellac & alcohol. Very tough.
5. Acrylics
6. Epoxys   These plastics can become hard upon the addition of a hardening agent. They have good adhesive qualities and resist heat, solvents and chemicals. Minimum shrinkage.

Epoxy resins are often used on surfaces for protection against the elements or oil and gasoline. These hard resins are not practical for use on flexing surfaces (such as small boat decks) as they tend to crack when flexed. Spar varnishes are more flexible and to date stand up better on flexing or warping surfaces.

7. Vinyls — Teflons (PAGE 63)
8. Electroplating — Cadmium, etc.

Chrome plating is really a combination of nickel and chrome and sometimes even a sublayer of copper. The underplating is often a layer of nickel on the metal or plastic part between .1 and 1.5 thousandths of an inch thick. The top chromium layer is usually only about 10 millionths of an inch thick but is a hard, abrasion-resistant, nonrusting surface. It can be bright or "brushed" for a softer satin look. Zinc, brass, steel or plastics can all be plated, although the zinc parts usually need a copper-plate base applied first. The nickel plating on items like toasters that receive limited wear may be as thin as .4 mils and gradually gets thicker on items like stools or golf shafts through patio furniture & bicycles to maybe 1.6 mils on items like car bumpers that really take a beating.

9. Cladding — Rolling the surface coating on with pressure. (Silver-clad copper, e.g.)
10. Hot Dip — In zinc, for example, as a galvanizing process.
11. Metal Spraying

12. Electrostatic Spraying

Paint is sprayed from a spray gun thru a device on the nozzle which creates an electrostatic field. The item to be painted is suspended from a ground or electrode. The particles of paint are thus attracted to ALL sides of the legs, braces, screw heads, etc., without the operator having to move the spray gun. A whole wrought iron chair, for example, can be painted in seconds with one burst from the gun.

13. Vacuum Laminating

The item to be coated one side is placed in a cavity and covered
with the particular laminating film. A vacuum is formed around the
item by pumping the air out of the cavity. Air pressure (15 pounds
per square inch, remember?) smashes the film into every contour of
the product and holds it there until the bonding is completed. Some
vinyls can be laminated to metal forms so tightly that the metal can
then be stamped, crimped, drawn or even welded carefully without
breaking down the vinyl laminate.

14. Veneering — Gluing or otherwise adhering a thin
material to surface, as in some formica
tops or cabinet veneers.

15. Plasma Spraying

Powdered paint is heated as it leaves the spray gun. The
molten particles then fuse into a paint film when they hit
the surface and cool. (Used on ceramic and metallic surfaces.)

16. Powdering

A film of dry powdered paint is dusted over a flat surface.
The item is then heated and the particles flow together as
they melt. (Wooden or thermoplastic products cannot be
heated in this manner.)

CASE HARDENING (Surface Hardening)

Treating the surface of a part to give it a tougher, more wear-resistant finish.
This can be accomplished by:

1. Heating to specific temperature for certain periods and then quenching in
liquids held at specified temperatures. The longer the "soak" or heating
time, the deeper the hardening. (See Section 4, Hardening.)

2. Carburizing — Introducing carbon into the surface layer
of low-carbon steels.

3. Nitriding — Introducing nitrogen into the surface layer.

4. Shot Peening — Blasting the surface of the part with shot seems to work
harden the surface and make it harder.

DIE CASTING

The young designer almost always has to seek advice from production men regarding methods of fabricating various parts of his design. Particularly in household, workshop and other appliance manufacture, the choice of metal alloy, plastic formula, or method of fabricating can be a very critical cost factor. Below are a few comparisons of materials and processes to give the student a rough idea of the many different factors involved in decisions of this type.

Plastic injection molding and compression molding have made considerable inroads into the older metal die-casting methods, but all the metal die-casting industry has increased its efficiency through research. Thinner walled castings, longer wearing dies, and faster cycling (production) rates are the result. The majority of appliance castings are of either zinc or aluminum, with magnesium and brass probably being no more than 5% of either zinc or aluminum castings. Iron is not used in die casting because of its high melting point (2800° F). The molten iron is so hot it destroys the surface of the die, so most iron is cast in sand molds or investment castings. Investment casting means that the mold is made out of a mixture of materials that is not affected by the 2800° metal but is somewhat finer and firmer than the usual sand molds. According to "Appliance Manufacturer," two firms in the United States have developed die materials that withstand the high temperatures necessary in ferrous die casting. Whether these developments have promise economically for stainless steel die casting and other applications remains to be seen.

Die casting firms and plastic molding plants (as well as forging, stamping or sand casting outfits) urge designers to consult with them DURING the design process. Production men in these fields can often lower cost by pointing out ways to simplify part dies, reduce material used, or increase strength without changing the integrity of the design. In one example I read about, a kitchen blender was designed in which 9 aluminum castings were reduced to 6, after consultation with the die-casting firm, with a resultant cost reduction of 12%. Young designers too often waste design time by not consulting early with production.

Below is a chart showing the general range of the characteristics of the four major die-casting alloys. A next step would be a more precise chart showing the exact characteristics of a SPECIFIC alloy such as SAE 903 Die Cast Zinc. Or in the plastic field it might be a chart showing the characteristics of a specific plastic such as a fire-resistant, glass-fibre reinforced silicone, etc. Such charts are more meaningful to production personnel, but the designer certainly should be educating himself as to the type of characteristics or specifications of various materials so that he can discuss his design somewhat intelligently with production men.

This chart contains very general information. It should be used by the student as a base for UNDERSTANDING what factors might enter into cost/function consideration — not as a production guide.

GENERAL COMPARISON OF DIE CASTING ALLOYS

	ZINC	ALUM.	MAG.	BRASS
ALLOY				
Melt temp. (F)	728°	1100°	1105°	1670°
Melt cost	1	2 x Zinc	2 x Zinc	4 x Zinc
Spec.grav.	6.6	2.7	1.8	8.3
lb/cu.in.	.25	.10	.07	.3
Cost/lb.	.85	.50	.80	1.20
Cost/cu.in.	.21	.05	.06	.36
Hardness	good	fair	fair	excellent
Machining cost	low	low	low	medium
Finishing cost	low	medium	high	low
Dimensional precision	excell.	good	excell.	fair
Corrosion resistance	fair	good	fair	excell.
Electrical conductivity	good	excell.	fair	good
Thermal conductivity	fair	excell.	fair	good
Part complexity	excell.	good	good	fair
Supply available	OK	OK	limited	OK
DIE				
Shots/hr (range)	200-550	60-200	75-300	50-300
Shots/hr (average)	400	150	225	150
Impact strength (rel.)	30	3	2	10
Min. thickness (in.)	.015 - .050	.04 - .08	.04 - .10	.055 - .090
Skin thickness (in.)	.012 - .024	.03	.03	.04
External threads (max. per in.)	32	24	24	10
Internal threads (max. per in.)	24	16	16	8
Die cost	low	medium	medium	high
Die life (shots)	500,000	125,000	150,000	50,000
FINISHING				
Chrome plating	excell.	fair	fair	excell.
Photo etching	excell.	excell.	excell.	good
Anodizing	good	excell.	fair	no
Lacquer - Epoxies	excell.	excell.	excell.	no

OTHER DESIGN FACTORS

1.   In metals the outer skin of the die-cast part cools more quickly and becomes dense and finely grained. The inner core of the part tends to become more porous and coarser grained. Thus strength, RELATIVE TO CROSS-SECTIONAL AREA, becomes less with increasing wall thickness. The designer should be aware that greater strength with less weight can often be achieved by using several thin-wall sections in place of massive walls to support loads.

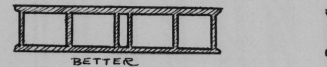
BETTER

SOLID

2.   The thicker the walls or higher the melt temperature the slower the production rate, because thick parts take longer to cool before they can be ejected from the die.

3.   External threads are easier to cast than internal threads. It is often cheaper to tap internal threads AFTER casting.

4.   Magnesium may be useful for designing low-inertia parts that need to be started and stopped quickly. Typical items are camera shutters, fans, calculator parts, distributor arms and lawn-mower housings.

The associations below are all concerned with researching of alloys. You might look up the local address of several of them and see if they have any brochures or periodical reports regarding new alloys, processes, or possible future breakthroughs in metal research that would give you some ideas for better design, economy of materials, or energy reduction.

ADCI	American Die Casting Institute
	366 Madison Ave., NY 10017
SDCE	Society of Die Casting Engineers
DCRF	Die Casting Research Foundation
ILZRO	International Lead Zinc Research Organization
INCRA	International Copper Research Association
MRC	Magnesium Research Center — Batelle Laboratories
	Columbus, Ohio
BNFMRA	British Non-Ferrous Metals Research Association

ASM   American Society for Metals, Metals Park, Ohio 44073 has a Reference Publications Department which has handbooks available on such subjects as: Properties & Selection; Heat Treating, Cleaning & Finishing; Machining; Forming; Forging & Casting; Welding & Brazing; and Aluminum.

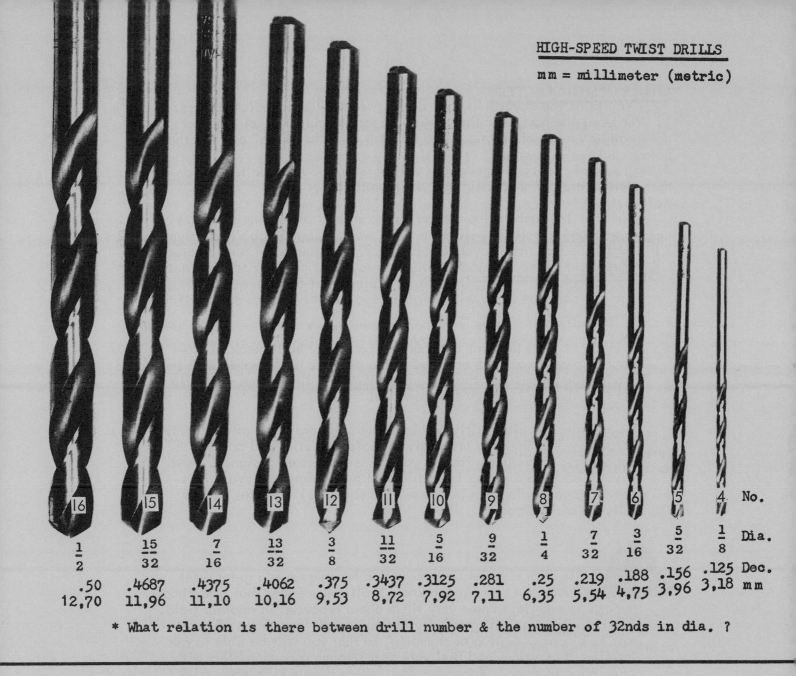

No.	16	15	14	13	12	11	10	9	8	7	6	5	4
Dia.	$\frac{1}{2}$	$\frac{15}{32}$	$\frac{7}{16}$	$\frac{13}{32}$	$\frac{3}{8}$	$\frac{11}{32}$	$\frac{5}{16}$	$\frac{9}{32}$	$\frac{1}{4}$	$\frac{7}{32}$	$\frac{3}{16}$	$\frac{5}{32}$	$\frac{1}{8}$
Dec.	.50	.4687	.4375	.4062	.375	.3437	.3125	.281	.25	.219	.188	.156	.125
mm	12,70	11,96	11,10	10,16	9,53	8,72	7,92	7,11	6,35	5,54	4,75	3,96	3,18

* What relation is there between drill number & the number of 32nds in dia. ?

PRODUCTION BUGS

Always remember that the production foreman knows a lot about his area. But also remember that you know a lot about your area. The majority of production men will respect your suggestions of form, placement, color, handle sizes and the 101 things the designer is hired to make decisions on. But HOW to get this design produced economically is a process problem, and there will be many "gray" areas where production's recommendations will have to be okayed, modified, or rejected. For example, let us say that the design is completed and you have specified an inch radius on the edges and corners of a casing cover for a small farm generator. The foreman points out that his shop machinery cannot handle larger than 1/4 in. bends. The inch radius will have to be done by an outside vendor and this will raise the cost on the casing by about $10 each.

You think the cover will look better with the larger radius corners. So you, the boss, and the production foreman usually get together with a variety of problems like this and work the "bugs" out together.

The total cost of the generator will be about $200. The ten extra dollars are 5% of that figure. Is it worth it?

Will the 1/4-inch radius corners be too sharp? After the generator is sold and installed, will the enamel wear off quickly on these sharp edges and make the cover look shoddy? Will the metal tend to crack when bent to the 1/4-inch radius? If so, what percentage of rejects would there be if doing it inplant? Rejects run costs up quickly. Or can the cover be made with thinner gage metal? And so on, until some conclusion is reached.

It might be well to point out that this particular bottleneck in production would not have occurred if the designer had cleared with the production foreman on the inplant equipment a little more carefully BEFORE the final specs were drawn up.

At any rate the nitty-gritty problems in design are quite often at this detail level in the gray area between sales esthetics and production economics; particularly for the young inexperienced designer. And if you also have a young inexperienced production man (Yes, it is possible, particularly in businesses just starting.) it makes for one jolly set of headaches.

Familiarity with a variety of production techniques is a great asset to a designer, but no man can know it all. Remember that oftentimes experts in one field know very little about processes in another field. You as a designer will gradually absorb enough knowledge of sheet metal, plastics, casting or stamping, so that you will be able to ask intelligent questions and feel your way through a cost problem with production foremen. Remember too, where the timid foreman will say, "Well I don't think we could do that!" A creative shop foreman will often say, "You design it, and we'll find a way to make it as economically as anyone."

All in all, however, whether you are young and inexperienced or a nationally known designer, there is a lot of fun in solving these design problems cooperatively. Respecting other men's backgrounds, listening thoughtfully to suggestions and arriving at solutions while still retaining the reins in your area of esthetics is a real challenge. Keep confident, keep cool, and try not to be too dogmatic at first. An industrial designer always has to consider economics as part of the design game. The best-looking farm generator in the world is of use to no one if it is priced out of the farmer's reach.

$ To Reduce Cost of Manufacture

1. Find and keep designer who can solve problems rapidly.

2. Find and keep the foreman who can encourage employees to reduce costs through self-analysis or shop-analysis of production methods.

3. Find and keep the draftsman who can interpret and draw rapidly and correctly.

4. Simplify design.

5. Use more standard pre-constructed units.

6. Use less material or less expensive material.

7. Use a more easily worked material.

8. Reduce number of operations in fabricating or assembling.

9. Use more jigged operations — thereby eliminating manual errors or operator fatigue.

    Jig System — A device to guide a tool in a certain path or a device to hold material in a certain position. Effect desired is usually mechanical and repetitive.

    Skill System — Effect desired is varied by operator.

10. Large firms can make time and motion studies of operations to cut down operator fatigue, simplify movements, or interrelate jobs. Production units can then be increased without pressuring employees to merely "Hurry up!" or raising piecework quotas.

But never forget that COST AS RELATED TO QUALITY is a primary factor in sales. Cost is meaningful only in relation to sales. The lowest cost item is the most expensive if it doesn't sell. The low bid may often be the costliest route you'll ever follow. It has been said that the biggest problem a manufacturer has is not making the products ...it's getting rid of them.

An excellent reference for the artist-designer who wishes to acquaint himself with basic engineering principles is a text titled "Design," Beakley and Chilton, 1974, Macmillan Publishing Co., Inc., New York, NY. As a product designer you will often be working hand-in-hand with the engineer. This reference will provide you with a good deal of understanding or 'interface' with the problems of the engineer.

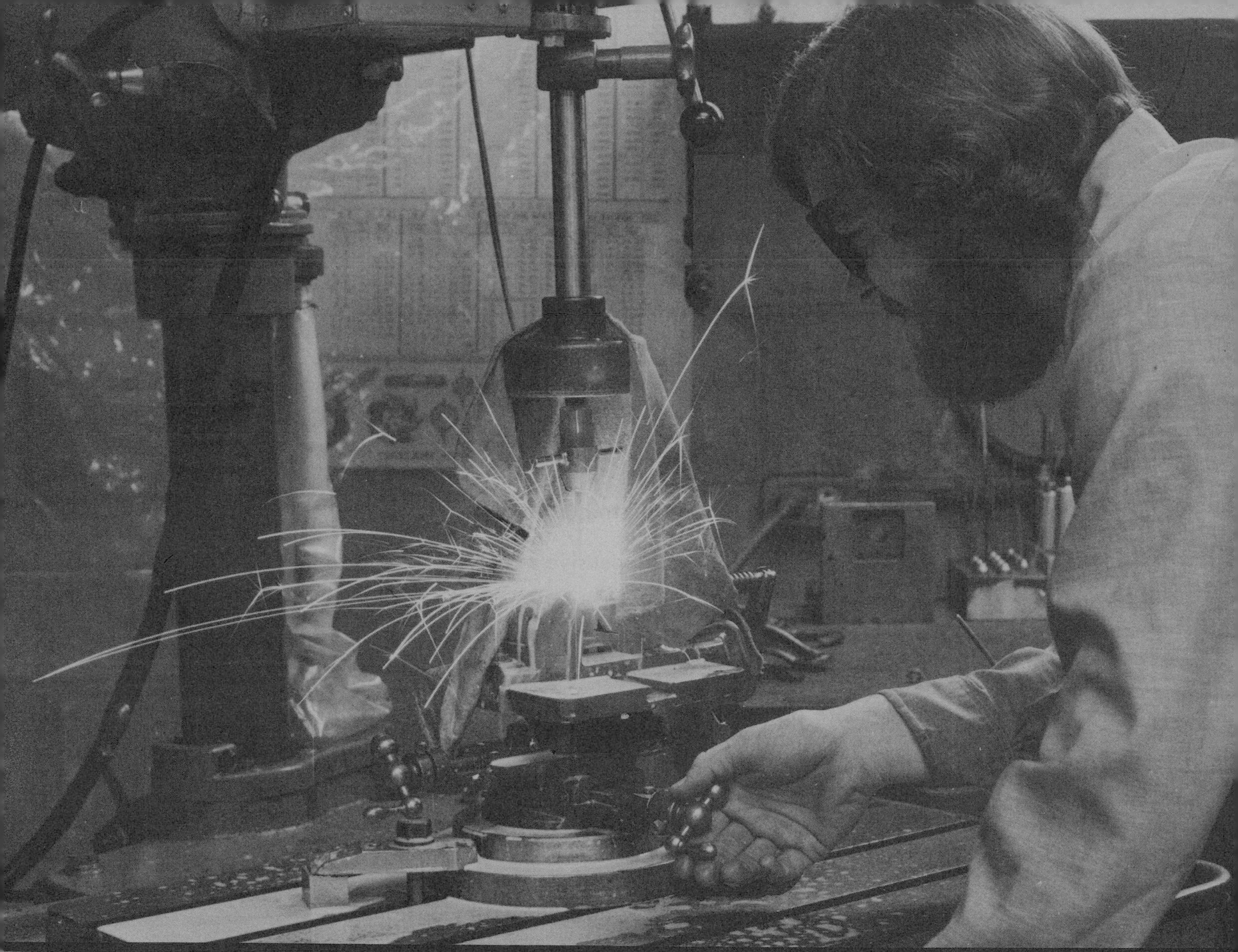

JOINTS & FASTENERS 6

ELECTRIC DISCHARGE MACHINE

Manufacturers production and assembly lines suffer from broken off studs, cracked taps and fractured drills stuck in holes. These components are often salvaged by sending them to a specialized machine shop which disintegrates the metal stuck in the hole without damaging the threads or the wall of the hole.

The part is grounded and a hollow vibrating electrode is thrust against the unwanted metal. The resulting arc gradually disintegrates the core. A coolant is introduced thru the center of the electrode which also flushes away the particles of metal. Voltage may be adjusted to size of job. Afterwards any metal left between threads or in flutes can be picked out.

Can you think of any other application for such a device?

Photo Courtesy of:
Helikon Products Co., Inc.
938 West Huntington Drive
Monrovia, CA.

The majority of beginning designers are concerned primarily with the esthetics of their products. This is fine, but they must also learn the business end of the profession. Market, cost of fabrication and assembly plus profit are an integral part of the design picture. Most things we use were designed to be mass produced. On the other hand, with the speeded-up assembly procedures, unit modules, and automated techniques almost anything could be mass produced, whether it was designed for the assembly line or not. However, the designer who understands fundamentals of assembly and can reduce his product into easily assembled modules can reduce cost. Some knowledge of a variety of joints, fasteners and movers is essential if the product is to be assembled economically. And often just as important is the case of disassembly for maintenance, repair or storage.

JOINERY

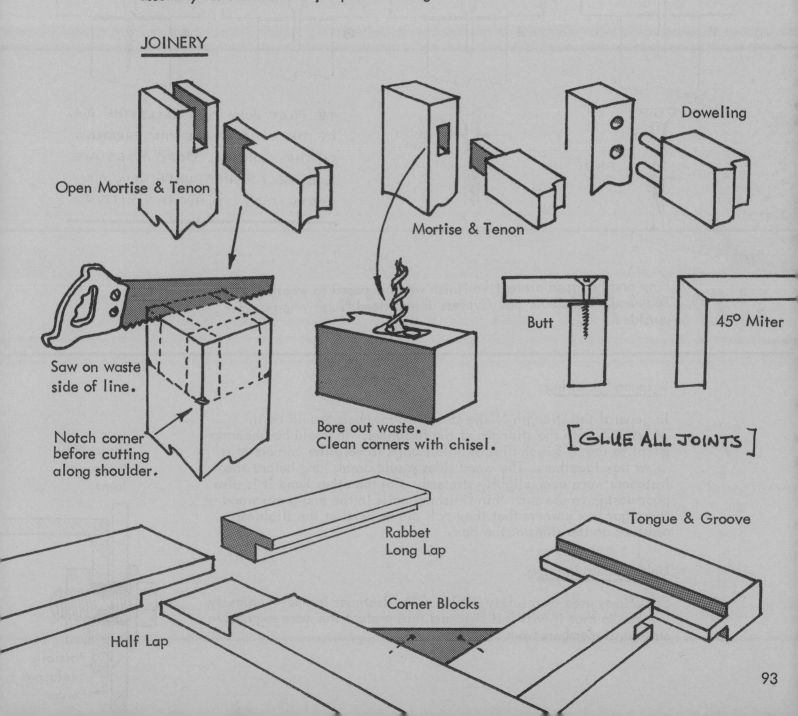

Open Mortise & Tenon

Mortise & Tenon

Doweling

Saw on waste side of line.

Notch corner before cutting along shoulder.

Bore out waste. Clean corners with chisel.

Butt

45° Miter

[GLUE ALL JOINTS]

Rabbet Long Lap

Half Lap

Corner Blocks

Tongue & Groove

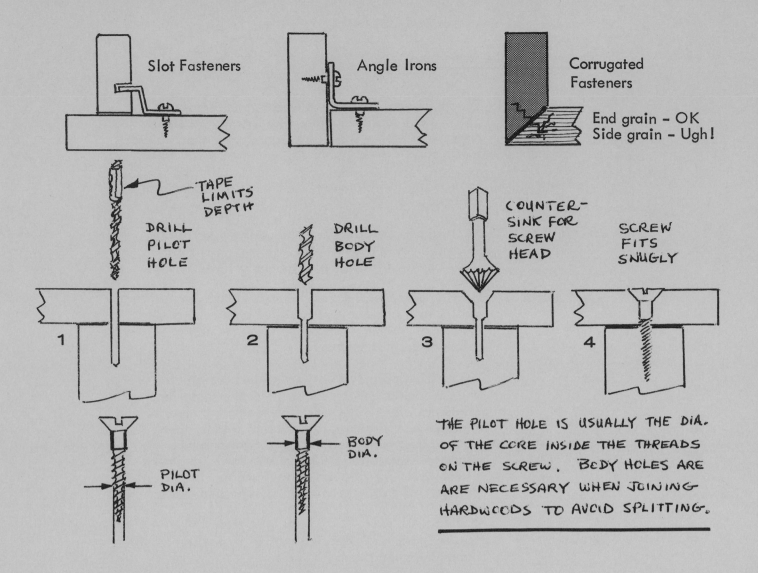

Slot Fasteners

Angle Irons

Corrugated Fasteners

End grain – OK
Side grain – Ugh!

TAPE LIMITS DEPTH

DRILL PILOT HOLE

DRILL BODY HOLE

COUNTER-SINK FOR SCREW HEAD

SCREW FITS SNUGLY

1   2   3   4

PILOT DIA.

BODY DIA.

THE PILOT HOLE IS USUALLY THE DIA. OF THE CORE INSIDE THE THREADS ON THE SCREW. BODY HOLES ARE ARE NECESSARY WHEN JOINING HARDWOODS TO AVOID SPLITTING.

Pine and Fir need protective finish when exposed to weather. Redwood, Red Cedar and Cypress do not need finish when used outside.

Balanced Jointing

In general the strength of the structural members should be in BALANCE with the strength of the fasteners. It would be uneconomical to use 1/8-inch dia. stainless bolts to hold the corners of a cigar box together. The wood sides would break long before the fasteners were even slightly stressed. On the other hand it is also poor design to use such thin finishing nails in the end-grain wood at the cigar box corners that they pull out or bend at the slightest pressure on the sides of the box.

Unbalanced Jointing

Sometimes used as a safety device. The fastener is made purposely weaker so that it will fail first and thus protect the more expensive structural members from distortion.

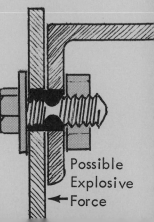

Explosion-Release Joint

Possible Explosive Force

Materials can be joined by:

Gluing Evaporative glues such as white glue, catalytic glues, heat-sensitive glues like Dry Mount Tissue, pressure-sensitive tape like Duct Tape, electrician's tape, etc.

Cladding Laminating thru pressure. Silver-clad copper, for example.

Soft Solder Heating parts to low heat and applying a fusible alloy such as a tin/lead solder which melts and flows between them (usually below 400°F).

Hard Solder Heating parts to high (red) heat and applying a similar fusible alloy (often silver solder) between them (from 400°F on up, depending on melting points of various solders).

The melting points of various solders are crucial when a component is very small or the soldering points are close together. The first point must be soldered with the high-temperature solder, the second point with the next lowest-temperature solder and the third point with the lowest-temperature solder. This prevents the first joint from melting apart when the second point is heated and so on.

Brazing Heating parts and applying very high melting solders between them.

Welding Melting the two parts at juncture. When cooled the metal is solidified as one piece. Welding all the way across a beam or frame member sometimes leaves an area of weakened metal just beside and parallel to the weld. Thus members are sometimes joined by welds only at intervals along the joint.

Pinching Crimping material around edge. Bending tabs over and flattening. Pinching material together and coating with hardener. Pinching clay together and firing to bisque.

Interlocking Mortising, dovetailing, buttons, zippers, slide locks, hooks, set screws, etc.

Penetrating Nails, screws, bolts, dowels, rivets, etc.

Tying Straps, buckles, laces, weaving, sewing, etc.

FASTENERS

Permanent (Requiring Permanent Bond)

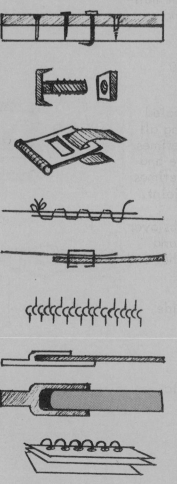

Welding

Spot Weld

Brazing

Soldering

Riveting (Hot)

Squeeze Riveting

Heat Sealing

Chemical Bonding

Ion Exchange Bonding

> A reaction eutectic solution is a material which when heated causes the two pieces of metal to flow together at lower temperatures than the regular melting point of either.

Temporary (Requiring Removable Fasteners)

Nails

Screws

T-Bolts

Bolts

Button

Clasp

Stitch

Lace

Staple

Zipper

Tongue & Slot

Slip Fit

Slip Groove

Spiral Spine

Nut Insert

Product designers often work with thin sheet metal which sometimes may require female thread inserts to accept machine screws. A typical form of insert is shown below.

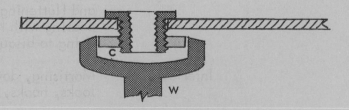

Nut insert provides female threads in thin material. Hand or automatic wrench (w) turns collar (c).

GULL ROCKER

Interesting use of slings, pins, and dowels. The function of the fastener becomes part of the visual esthetics.

Design Patent No. 210,939
Designer: George Schwarz

Courtesy Gold Medal Folding Furniture Co., Racine, Wisc.

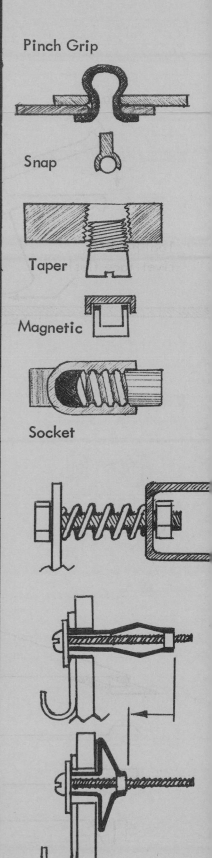

Pinch Grip

Snap

Taper

Magnetic

Socket

Spring Load

A bolt with a helical spring over its shaft between the two parts allows for pre-set play between the parts. Sometimes used where vibration damping is necessary between faces. Spring specifications can be changed depending on vibration.

Blind Anchors

Blind anchors or blind bolts are any device that can be torqued into place without having access to one end of the bolt. This can be accomplished by having a free-turning machine screw or bolt working into a fixed nut, as in the diagram below. As the nut advances along the bolt, it crimps or bends two pieces of metal outward from the shaft which eventually binds the two pieces together. Thus the hole can be drilled thru the two pieces of material, the blind bolt can be inserted through the hole, and the two pieces of crimped metal act as a washer to prevent the nut from coming back out through the hole. Pretty nifty, huh? There are other variations. A sort of butterfly nut may be inserted also, which springs open once it is pushed through the hole and acts as the retaining washer.

97

Pop Rivets

These are a type of light aluminum or other alloy rivets which are literally turned inside out as a shaft with a knob on the end is pulled through its center. (See diagram below.) These are used for light applications only but are very efficient when the "riveting tool" is used with the right pressure and firmness. The tool uses a lever action through a long fulcrum and a pliers-like jaw to pull the shaft through the rivet hollow core and pop it out on the other side.

Squeeze.

Tool holds
rivet face flush.

Jaw pulls shaft up.

After crimp, shaft breaks off.

Knob crimps rivet.

Interlocks

Sheet metal and plastic can be formed or crimped in a variety of ways to interlock. Usually one side is male and the other side of the part a matching female fit. Sealants such as silicone can be fabricated with the part or applied when fitting panels together. A variety of textures can be achieved by the shade and shadow of the contoured joints. Aluminum house siding, house-trailer paneling, and product cases or metal business desks are typical applications. A few cross-section examples below.

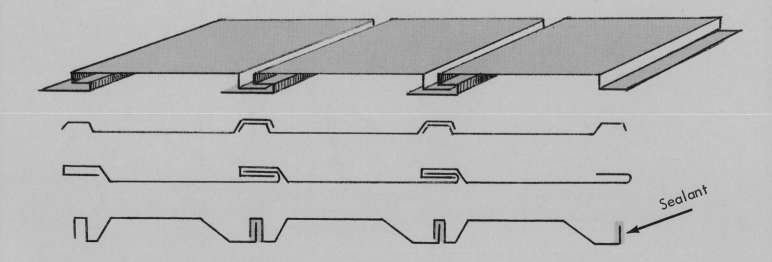

Sealant

Gas Welding

In gas welding, acetylene is burned at the end of a nozzle with oxygen, which make a very hot flame. This flame is washed over the two pieces of metal to be joined until they become almost molten (cherry red). Then a welding rod is introduced into the flame which melts and combines with the two pieces of metal.

A flux may also be introduced into the joint. The flux attracts impurities and keeps them from polluting the weld. After cooling, the joint is cleaned with solvents and wire brush. The preparation of a joint in welding is of utmost importance. The joined surfaces must be as clean as possible to prevent any foreign material from weakening the joint. This is usually accomplished by grinding, wiping, or using solvents.

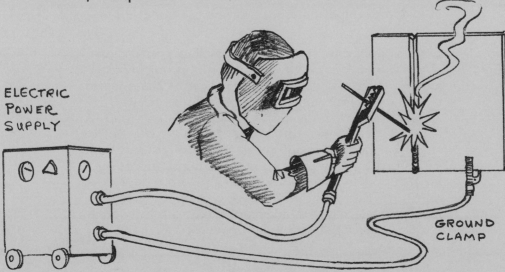

Arc Welding

In arc welding, the parts to be welded are connected to a clamp which is grounded to the power supply. The other lead from the power supply goes to the hand clamp which holds a long, thin electrode. The electrode is merely another welding rod covered with a flux coating. When the electrode is touched to the grounded parts it sparks, and electricity flows thru the electrode into the joint, creating a high temperature which melts the electrode and heats up the joint.

The welder controls the distance of the electrode from the parts: the closer, the hotter; the farther away, the cooler; and thus controls the puddling of molten metal as the electrode moves along the seam leaving the cooling weld behind it. And it ain't easy, Mac. It takes a lot of skill and years of practice to produce strong welds free of slag with minimum distortion of joints without overheating the parts so much that the properties of the original metal are affected. There are automatic welding machines that work by feeding in a long coil of wire, but this is applicable only when standard seams can be jigged up for repetitive operation.

Resistance Welding

Resistance welding is often used for spot welding material together. Two electrodes are placed opposite each other and pressed together with the metal joint between them. The electrodes are given a surge of electricity and the metal between resists the current, becomes hot, and welds itself together in that one spot.

Resistance-Welded Fasteners

These fasteners have bits of metal attached which weld the fastener to a surface when electric current is passed thru the part.

Thru Hole   Rt.-Angle Stud   Rt.-Angle Bracket

The low-carbon steels are more easily welded than complex alloys. In fact, dissimilar metals are very difficult to weld together and are usually joined by soldering or brazing. Soldering, in general, is done at low heats, under 400°.

Brazing   Some high-carbon steels, stainless steels, and many of the nonferrous metals are difficult, if not impossible, to weld. Brazing is a method of joining these metals or two dissimilar metals by using a nonferrous alloy (solder) that melts around or somewhat above 800 degrees. This temperature must be below the melting points of the metals being joined. Many steels don't melt until 1500 degrees, so there is considerable latitude. The two pieces to be joined are heated up to the melting point of the fusing alloy, which then flows between the surfaces and makes a joint weaker than a weld but stronger than adhesives.

NAILS

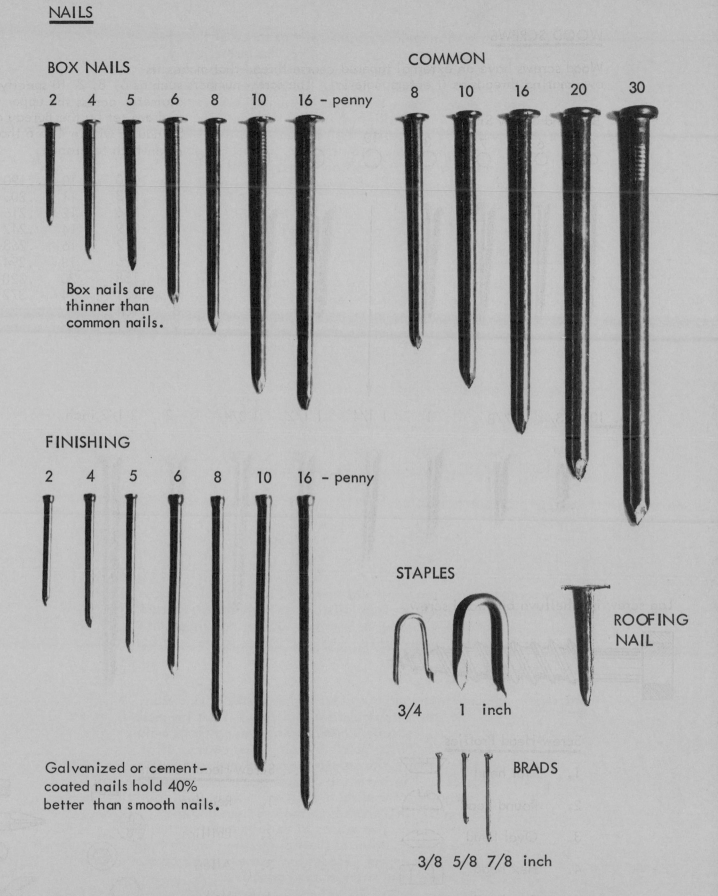

BOX NAILS

2   4   5   6   8   10   16 – penny

Box nails are
thinner than
common nails.

COMMON

8   10   16   20   30

FINISHING

2   4   5   6   8   10   16 – penny

Galvanized or cement-
coated nails hold 40%
better than smooth nails.

STAPLES

3/4   1   inch

ROOFING NAIL

BRADS

3/8 5/8 7/8 inch

Nails in olde England were sold in lots of 100. If a hundred nails cost 10 pennies,
that size nail was called a 10-penny nail. The smaller the nail the cheaper the price.
Today the nails are standardized but have retained the old numbers as size designation.

WOOD SCREWS

Wood screws have an external tapered coarse thread that makes its own mating threads as it enters material. The screw numbers such as 6, 8, & 10 specify the diameter across the upper shank and are set by the Bureau of Standards within 4 to 6 thousandths of an inch tolerance.

Bolt and Screw Sizes

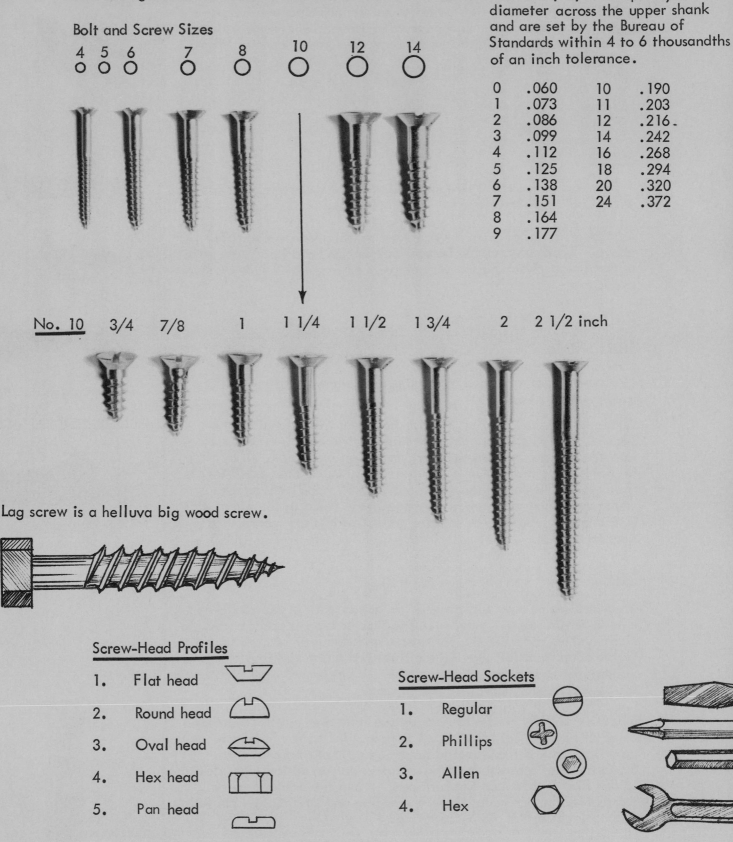

4 5 6 7 8 10 12 14

0	.060	10	.190
1	.073	11	.203
2	.086	12	.216
3	.099	14	.242
4	.112	16	.268
5	.125	18	.294
6	.138	20	.320
7	.151	24	.372
8	.164		
9	.177		

No. 10   3/4   7/8   1   1 1/4   1 1/2   1 3/4   2   2 1/2 inch

Lag screw is a helluva big wood screw.

Screw-Head Profiles

1. Flat head
2. Round head
3. Oval head
4. Hex head
5. Pan head

Screw-Head Sockets

1. Regular
2. Phillips
3. Allen
4. Hex

102

METAL SCREWS

Machine Screws (Cap Screws)

Machine screws look like bolts but screw into a tapped hole in a part. (Bolts are designed for use with a nut.) There are many varieties of screws, but a few common ones have heads as follow:

Hex   Flat   Allen Socket   Round   Oval   Pan

Fillister — EXTRA STRONG

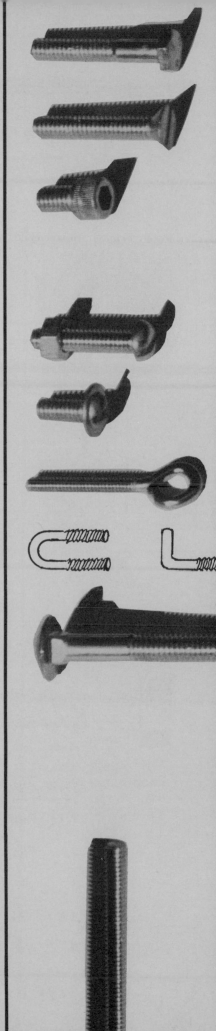

BOLTS

A bolt is an externally threaded rod with head designed for use with a nut. When torqued it holds assembled parts together.

Bolt heads are similar in shape to the machine screw examples above. In some cases the bottom side of the head has a thin slice of metal left to simulate a washer (washer faced). A few other heads are:

Eye Bolt   U-Bolt   Angle Bolt

Stove bolts are long, skinny bolts where shear strength is not a critical requirement. They range roughly from 1/8 to 1/2 inch in diameter and from 3/8 to 6 inches in length. Their heads are usually flat or round.

Carriage bolts are similar to stove bolts, but the underside of the head is square, so that when the head of the bolt is driven into wood it will not turn as nut is tightened on other end. They vary in diameter from 3/16 inch on up and may be as long as 20 inches in length. They were used to bind wood to wood, or wood to metal over long spans when carriages were made in the olden days. Don't laugh, you may be looking for a few of them yourself in days to come.

STUDS

Studs are metal rods with threads on one or both ends, used when a long tension truss is needed. The shorter type is used in applications such as a cylinder block where one end may be sweated* or welded into position leaving the threaded end protruding. Has advantage of long thread distance, which is important when tightening space varies with gasket thickness, etc.

*Sweating usually means to heat up the hole area, let's say, then drive or screw a cold stud into the hole. When the hole cools it contracts around the stud and holds it in place permanently. Sweating also refers to heating metal to partial fusion and letting solder flow in between the surfaces. When the colder stud or solder hits the hot metal there is smoking and steaming, from which probably comes the word, sweating.

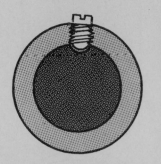

Tapping Screws

Screws that cut a thread into the material hole as they go in. Thus pretapped holes are not necessary. These screws can also be case hardened and driven in very tightly.

← Cutting edge which cuts mating thread.

← Chip cavity.

Set Screws

Semi-permanent compression clamping device to hold collar-like parts from rotating or slipping on shafts.

Set screw diameter recommended is 1/3 to 1/4 shaft diameter. However, the majority of set screws today are too small and wear out or fail to hold in a multitude of applications. A predrilled dimple on shaft (see drawing) helps set. Set screws usually should be harder (Rockwell Scale) than the shaft and, if critical, case hardened.

Case Hardened   A heat treatment method of making the outside surface (case) of the metal hard, but leaving the interior tough and ductile.

Set screws have points that are:

| Round | Conical | Flat | Dog | Cup |

NUTS

A nut is a piece of material having an internal thread which mates with a bolt. A few varieties are:

Hex   Cap   Wing   Castle
(Used with cotter pin)

Sleeve

Often used as a smooth bearing or axel for rotating members.

Locknuts

A jamming or wedging action prevents easy removal. Accomplished by tapering, distorting threads, out of round, double nut pressure, cotter pins, etc.

Spring-loaded lockwashers that cut on back-off should not be used on soft materials.

This is a stamped out, single-thread, light-duty, spring-action locknut that is inexpensive.

GASKETS and SEALS

1. Gaskets

A gasket is usually a flat, resilient piece of material used as packing between two interfacing surfaces. The gasket between the cylinder block of your auto engine and the head is a typical application. The gasket is often coated with a varnish-like substance and acts as a seal by being compressed as the head bolts are tightened.

2. Sealing Washers and O-Rings

Washers are in a sense small gaskets. The washer in the nozzle of your garden hose prevents water from squirting out between the two metal interfaces.

O-Rings are washers with a circular cross section. When compressed between two surfaces they are flattened and often give a more efficient seal than a flat washer.

Retaining Rings

There are many variations of retaining rings, but essentially they are a ring of material which snaps in place next to the seal or part to prevent it from moving.

Hermetic Seal — To make airtight by fusion or sealing. However, in applications where air leakage is critical to the operation of the instrument, seals rarely meet rigid specifications. Therefore, welding metal, fusing glass, or hard soldering are usually the best solution.

3. Boots

A boot is a cylindrical, elastic piece of rubber or plastic that is slipped over a joint to keep dirt out and grease in.

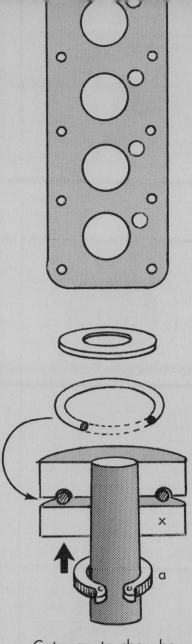

Cutaway to show how O-ring is held in place by annular groove.

Retaining ring (a) is slipped up on shaft to hold part x against O-ring

Boot

ADHESIVES

In general, there are about 100 different chemical types of adhesives packaged under approximately 1000 brand names. The following general classifications may simplify understanding for the student designer and model maker.

Rubber & Contact Cements

If applied to both surfaces and let dry before joining, the bond is instantaneous. Good for nonporous surfaces such as tempered masonite because there are no solvents in the glue that have to seep out to the edges to evaporate.

Mastic

Thick pastes. Fast solvent evaporation. Good for use on uneven surfaces. For example, acoustical ceiling tile can be laid very evenly against a slightly uneven subceiling as the gook squishes into the hollow areas.

Epoxy Resins

Catalytic reaction between resin and hardener gives tough bond. Cures fast on hot days, slow on cold days. Curing stops completely at $0°$ F. Adhesive may set up flexible or brittle depending on type or amount of hardener used. Often comes in two containers which are mixed together before applying.

White Glues, Vinyl Emulsions

Evaporation of water starts curing. Usually take 24 hours for set. Good on wood, paper and fabric. If diluted with water can be used for primer on old concrete surfaces which must accept new concrete.

Hide, Animal or Vegetable Glues

Low-strength adhesives for inexpensive use indoors. Library or classroom pastes are examples. Bond decreases if area gets wet.

Hot-Melt Glue Guns

A thermoplastic stick of glue shoved thru a trigger-quick heating nozzle on a glue gun will give a squirt of quick-bonding adhesive for fast action on small jobs. Glue sets when cool.

Silicones

Combine with moisture in air to form a tough, rubbery solid. Cures slowly below 30% humidity. Will cure under water. Resists heat to $500°$ F. Stays flexible to $40°$ below zero.

Note: Before you use a new adhesive, try a test sample !

There is a very comprehensive article on adhesives titled, "All You Need to Know about Adhesives," by R.C. Snogren, in the magazine "Popular Science," December, 1971.

Plastic Bonds

a. A variety of chemical liquids used to bond plastic sheets together. Surface must be clean. Slight abrasion of surface helps bond. Hypodermic needle or needle-like caps on glue bottle can be used to run a bead of liquid along butted joints. Liquid seeps between bonding surfaces by capillary action.

b. A solvent that attacks plastic can be used to soften the bonding surfaces. Then bring edges together with a slight clamp pressure. Protect other surface areas by masking off before apply-ing solvent. (Acetone, Ethylene Dichloride are examples.)

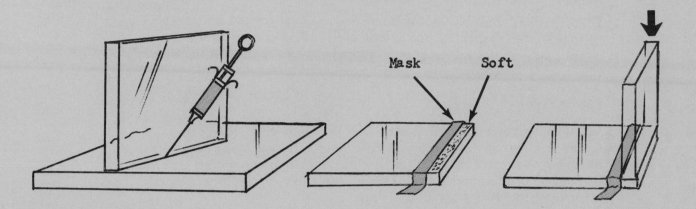

Mask   Soft

FORCES (Fasteners and Function)

Forces to be considered by the designer :

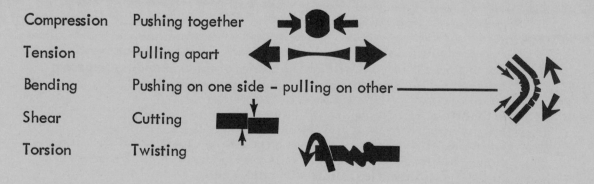

Compression	Pushing together
Tension	Pulling apart
Bending	Pushing on one side - pulling on other
Shear	Cutting
Torsion	Twisting

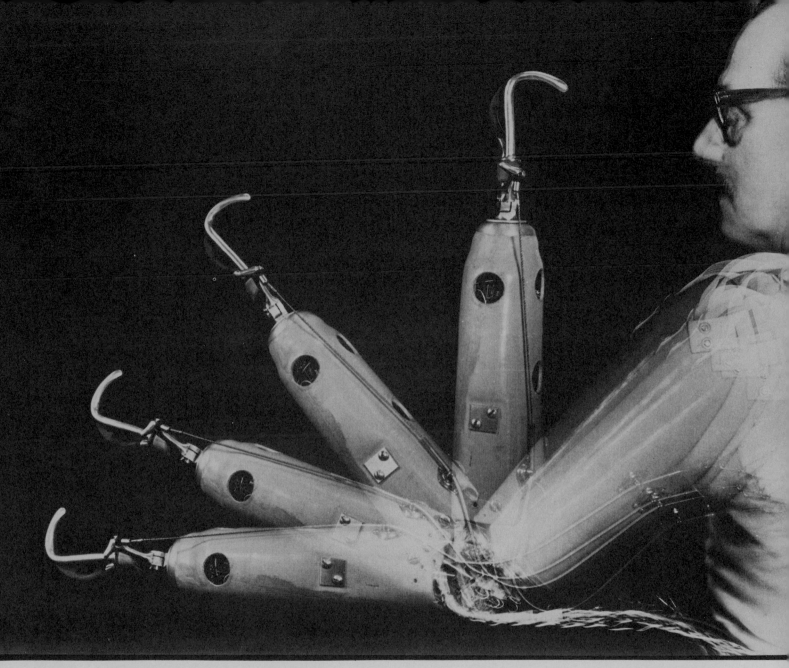

7 MOVERS and CONTROLS

BOSTON ARM

Artificial arm employs natural electric signals from muscles in amputee's stump. In other words it is an arm that the amputee can THINK into action. When the amputee wills a nonexistent arm into action the stump muscle gives off an electrical signal as it contracts. This tiny signal is amplified and used to control a battery-powered electric motor in the arm.

Joint developers with Massachusetts General Hospital, Boston, 1968, were Massachusetts Institute of Technology, Harvard Medical School and Liberty Mutual Insurance Co.

Could you design a circuit so that a person could think-operate a simple machine?

Photo courtesy of:
Massachusetts General Hospital
Boston, Mass.

SECTION 7   MOVERS & CONTROLS

Movement creates dynamics known as:

1.   Acceleration   Change of velocity or rate of such change.

2.   Force   Any action (such as pushing or pulling as examples of mechanical force, or thermal, electrical, chemical, or magnetic, etc.) which changes the relative condition of two bodies as to rest or motion.

3.   Shock   A big Bang.

4.   Vibration   A series of little bangs.   ***)))

5.   Pressure   Squeezing.

    a.   Liquids –   Hydraulic   Both are fluids.
    b.   Gases –   Pneumatic

MOVERS

1.   Electric Motors

    A series of revolving electromagnetic armatures pulling against permanent magnets timed by a simple distributor. Size can range from those the size of a pin head to those as large as eight feet high. Dynamos run off of water turbines might be included in this category.

2.   Bionic Motors

    A rather recent development. These small motors get their energy from the electrical impulses of the human body that excite and control muscles. Even though the signal is weak and the motor output is extremely small, these bionics could control larger motors. The implication is: Can we initiate force or do work merely by thinking about a specific action?

3.   Thermocouples

    Two dissimilar conductors side by side subjected to heat variation give off an electric current. Thermoelectric generators range from 1/2 to 1000 watts.

4.   Turbines

    Essentially a propeller system of many blades. Steam, water, or heated air can be the driving force.

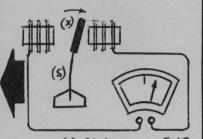

MASS (X) BALANCED ON FLAT
SPRING (S) FALLS BACK AS
AIRCRAFT ACCELERATES.
FIELD BETWEEN COILS
ACTIVATES READOUT.

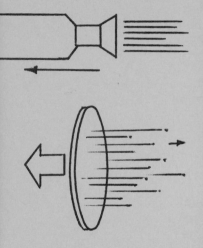

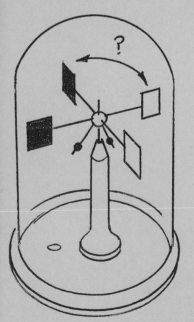

5. Internal Combustion

Explosions are used to drive pistons which transfer the force through a connecting rod to a crankshaft which rotates on bearings. "Internal" means that these explosions, whether from gasoline vapor, Diesel fuel, natural gas or otherwise, must be housed in a heavy case or engine block.

6. Springs

Spiral springs in clocks, coil springs as in screen door checks, flat springs in accelerometers all provide a force or motion after being "wound up."

7. Rubber Bands

After being stretched, will "recoil."

8. Jet Propulsion

Squirting gases or liquids out in one direction to gain motion in the opposite direction through reaction. (For every force there is an equal and opposite reaction.) Turbines in series are often used to increase force of ejection.

9. Ion Propulsion

Squirting ions away from a vehicle in space gives it a small impetus. But if this impetus is kept up over a long period of time it can mean increasing acceleration through accumulative bits that could thrust the vehicle to infinitesimal speeds, as there is no retarding friction or gravity in outer space.

10. Light Propulsion

Make a set of four balanced vanes, shiny on one side, black on the other. Balance it delicately on a pinpoint bearing within an evacuated glass globe. Set it in sunlight and the vanes will slowly revolve. The light "particles" or "photons," or whatever they are, bounce off the reflective surface and thus provide a reaction of pushing the vane away. The light rays striking the black side are absorbed into the black "sink" and don't get a chance to bounce. Net result — motion.

If the globe is not evacuated, the vanes will turn the other way. The black surfaces heat up faster than the shiny surfaces, because they absorb more light energy. The air next to these black sides becomes hotter, thus the air molecules are excited to greater activity. They bombard the black side of the vanes with greater force than they do the shiny sides. So-o-o the black side is pushed away.

Which of the above examples do you believe is true?

11. Wind

Prevailing winds, gusts, thermal updrafts, canyon downdrafts can all provide force for gliders, sails, windmills, wind turbines, or wind rotors.

12. Solar Energy

Heat can be gathered from the sun's rays by a "sink," such as large black panels which absorb the heat and transfer it to circulating water in a pipe system. Or the heat can be gathered by shiny panels reflecting the rays to various size focus centers. These rays coming off a parabolic reflector can drill holes in metal, or act as a solar furnace to convert water into steam. More spherical reflectors can provide heat for cooking or other general heating purposes with less danger, because the rays are not as concentrated.

There are solar conversion cells, such as selenium or silicon units, which harness the sunshine and convert it to an electrical output. NASA Cleveland Research Center at Cleveland, Ohio, is doing research on these solar conversion units. A variety of solar energy publications are available at Zomeworks, 1212 Edith Blvd., Albuquerque, New Mexico 87103. Federal government bureaus may have information also.

13. Gravity

One of the greatest scientific puzzles of all time. If man can ever direct it or nullify it by shielding we will probably have all the energy we will ever need. Water gives us energy by being higher in one spot than another. Electricity gives us power by having a high potential in one "box" and a lower level in another, so it flows also. Even an explosion is the movement (albeit damn fast) of particles from one place to another because of different "levels." Gravity shielding would allow us to raise a weight with less energy than we dropped it. "Perpetual" motion would almost become a reality, and we wouldn't have to use up our world by exploding, or burning it. Then, of course, we'd have the greater problem of over-shielding and nudging our earth out of orbit. Hm-m.

14. Biology

Muscles constrict upon signal from minute amounts of chemicals released by nerve impulses. Cells multiply and things grow larger. Slowly, yes, but this is also motion. Conceivably, if speed of cell growth could be fantastically increased and controlled, initiated and released in some manner, then one would have a nonending source of power also. Whereas cancer, at present, is a disease we are trying to control, we may find that runaway growth in some types of cancerous cells may be just what we are looking for as the key to power from growth, rather than power from destruction. Don't hold your breath ... but how would you like to tie into that one for a Ph.D. thesis?

OTHER ENERGY SOURCES

Much of the information below is condensed from a six-page report entitled, "Solutions to the Energy Crisis," available from the Environmental Alert Group, 1543 N. Martel Ave., Los Angeles, CA 90046, U.S.A.

1. Placing turbines in fast-moving ocean currents such as the Gulf Stream which moves between 3 to 6 miles per hour.

2. Waves could force water through a one-way valve system into a reservoir above sea level. Returning water could run turbine. Floats attached to long booms could ride up and down over waves and turn a generator.

3. In a few places tides can flow in behind dams every 6 1/2 hours.

4. Sea Thermal

   The warm water at surface of ocean and the colder water 1000 feet down can vary as much as 45 degrees F. Heat engines might be operated by tapping this differential with condensers and long, two-way flow pipes reaching to the sea bottom.

5. Geothermal

   This includes hot water and steam geysers, but hot rock just below the earth's crust seems to offer the greatest potential. A Downhole Heat Exchanger, designed by Van Huisen, is a closed system of long, double-walled pipes inserted into the hot rock. Water is pumped down the inner pipe and returns up through the outer pipe as steam. As the circulating water is in a closed system there is no pollution either below or above ground of escaping sulphur fumes, hydrogen sulphide, ammonia, or highly mineralized waste waters, which often cause problems with open systems.

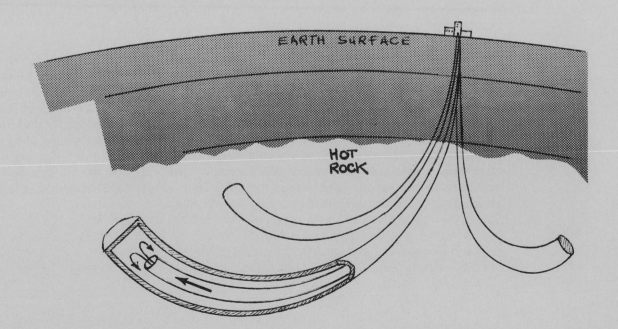

6. Methane

    Can be produced from organic wastes, garbage or algae. It is estimated each pound of organic waste could yield about 10 cubic feet of methane during anaerobic digestion (bacterial decay in the absence of air).

7. Pyrolysis

    A similar decaying process which can produce crude oil.

8. Nuclear

    The big problem here still remains: What to do with the nonending amounts of dirty, radioactive wastes that come   not so much from the nuclear power plants themselves, as from the plants that MANUFACTURE or "breed" the fuel for the power plants. A few years ago there were several articles condemning the methods of storing radioactive wastes at Hanford, Washington. The criticism seemed to center on the issue of whether the stainless steel tanks that were holding the liquid wastes were monitored well enough to keep them from corroding out and depositing radioactive materials in the water table. Or did the government have the right to dump them into the ocean, etc.? At that time efforts were being made to change the processing of fuel so that the radioactive wastes would be in granular form, rather than liquid, which theoretically would be easier to handle and contain.

9. Contained Fusion

    The Energy Research & Development Administration (ERDA), which is the new title for the Atomic Energy Commission has been doing research regarding energy release from Deuterium (a heavy gaseous isotope of hydrogen) bombarded by Laser beams. This is only one of several projects which, if successful, would mean atomic energy without radioactive wastes. Successful models have been made, and now a larger facility is being built to test the practical nature of it.

Whether successful or not, all these various possibilities are something designers should keep cognizant of. Trends of this nature, which make possible safe (controlled or contained) units of energy, will certainly have a tremendous impact upon the manufacturing process, the size of tools and machines, as well as cultural changes. Also, I feel it is encouraging to know that the same commission that has so often been criticised, is now the source of research to make the energy release safe and clean; just as the Jet Propulsion Lab of the California Institute of Technology, (which built the Ranger Space-Craft sent to the moon) is now using part of its budget, facilities and scientists to do ecological research.

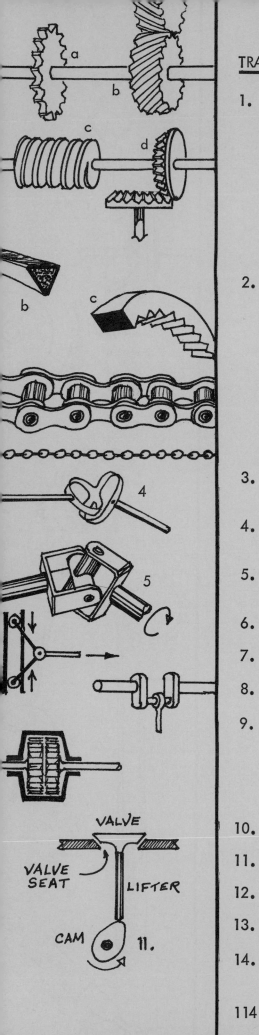

TRANSMISSIONS

1. Gears
   a. Spur gears – Teeth parallel to shaft
   b. Helical – Teeth follow helical pattern
   c. Worm – Spiral gear for power
   d. Bevel – Shafts intersect at angle

   Gear teeth are shaped so the second tooth accepts the load before the first tooth finishes its push. This gives smoother power. Helical gears (teeth set in a slight curvilinear pattern) give even greater smoothness.

2. Belts
   a. Flat – Used on conveyor systems or on wide drum wheels.
   b. V-belt – Prevents side slip.
   c. Timing (toothed) – Prevents rotary slippage.
   d. Roller bearing chain – Heavy-duty applications.
   e. Bead chain – Lightweight applications only.

3. Flexible shafts – For transfer of rotary power between points which cannot be joined by solid shafts.

4. Flexible couplings – Change angle of rotary motion in light-duty applications.

5. Universal joints – Change angle of rotary motion in heavy-duty applications.

6. Ball & socket – Lever action, not rotary.

7. Toggle joints – Produce motion at right angle to thrust.

8. Crank shafts – Change reciprocating action to rotary (Vice Versa).

9. Automatic transmissions –

   A multibladed turbine wheel placed face to face, close to but not touching, a second turbine wheel in a cavity filled with oil. As one wheel is rotated, the oil starts in motion and acts on the second wheel.

10. Pulleys & sheaves – Vary direction and power.

11. Cams – Change rotary action to reciprocating directional action.

12. Wheels – You name it.

13. Propellers – Rotary action to thrust (Vice Versa).

14. Hoses – Contain hydraulic or pneumatic pressure.

114

CONTROLS

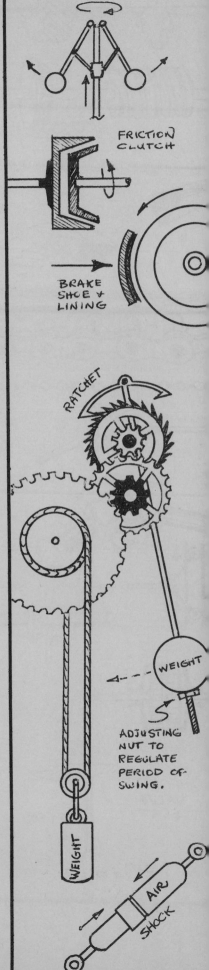

1. Governors

   Automatic attachments for controlling machine speed, usually by regulating fuel supply. Often operate through centrifugal action of two balls which activate a throttle linkage. Other devices can control gas or liquid flow, etc.

2. Clutches

   Control transmission of energy through discs, bands or shoe friction, hydraulic turbines, etc.

3. Brakes

   Control dissipation of energy through discs, bands or shoe friction, etc.

4. Flywheels

   A heavy wheel which, through its inertia, moderates any unwanted fluctuations in machinery speed. May also act as a gyroscope to maintain steady axis within a gimbal; or moderate pitch, yaw or roll in a vessel.

5. Pendulums

   A body suspended from a fixed point that swings to and fro under the impetus of gravity and momentum. Often used as clock regulators. Torsion pendulums twist clockwise and counterclockwise instead.

6. Ratchets

   Reciprocating or stationary pawls which engage teeth on a wheel to turn it forward or keep it from turning backward. Used to time clock movements or regulate intermittent movements in machinery.

7. Rudders

   Control direction by flow of fluid past its surface.

8. Valves (See previous page.)

   Devices for closing or modifying the flow of fluids through pipes, inlets, etc. Musical instrument valves, as in a trumpet, are calibrated to change harmonics in relation to air flow.

9. Dampeners

   Check or retard action. Examples are shock absorbers or even simple rubber mounts.

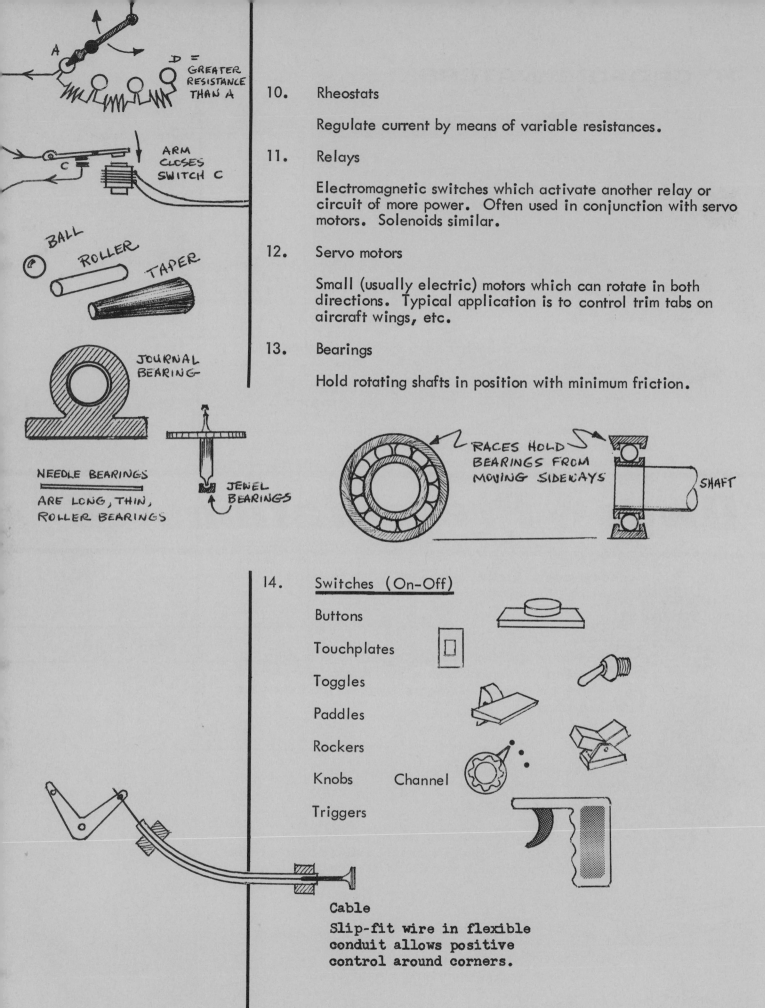

A

D = GREATER RESISTANCE THAN A

ARM CLOSES SWITCH C

C

BALL
ROLLER
TAPER

JOURNAL BEARING

NEEDLE BEARINGS
ARE LONG, THIN, ROLLER BEARINGS

JEWEL BEARINGS

10. Rheostats

Regulate current by means of variable resistances.

11. Relays

Electromagnetic switches which activate another relay or circuit of more power. Often used in conjunction with servo motors. Solenoids similar.

12. Servo motors

Small (usually electric) motors which can rotate in both directions. Typical application is to control trim tabs on aircraft wings, etc.

13. Bearings

Hold rotating shafts in position with minimum friction.

RACES HOLD BEARINGS FROM MOVING SIDEWAYS

SHAFT

14. <u>Switches (On-Off)</u>

Buttons

Touchplates

Toggles

Paddles

Rockers

Knobs   Channel

Triggers

Cable
Slip-fit wire in flexible conduit allows positive control around corners.

HYDROFOIL WATERSKI

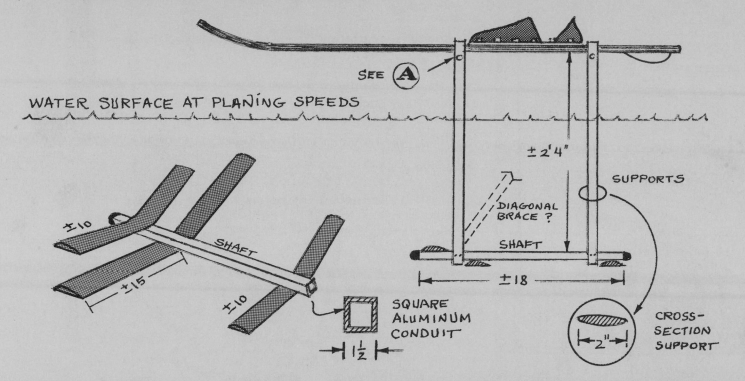

WATER SURFACE AT PLANING SPEEDS

SEE Ⓐ

±2'4"

DIAGONAL BRACE ?

SHAFT

SUPPORTS

±18

CROSS-SECTION SUPPORT

2"

±10

±15

±10

SHAFT

SQUARE ALUMINUM CONDUIT

1½

Dwgs & Presentations   8

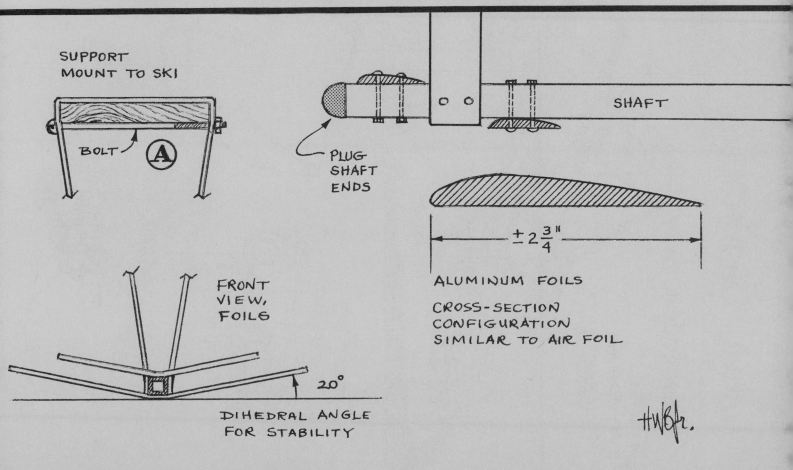

SUPPORT MOUNT TO SKI

BOLT   Ⓐ

PLUG SHAFT ENDS

SHAFT

±2¾"

ALUMINUM FOILS

CROSS-SECTION CONFIGURATION SIMILAR TO AIR FOIL

FRONT VIEW, FOILS

20°

DIHEDRAL ANGLE FOR STABILITY

HWBfr.

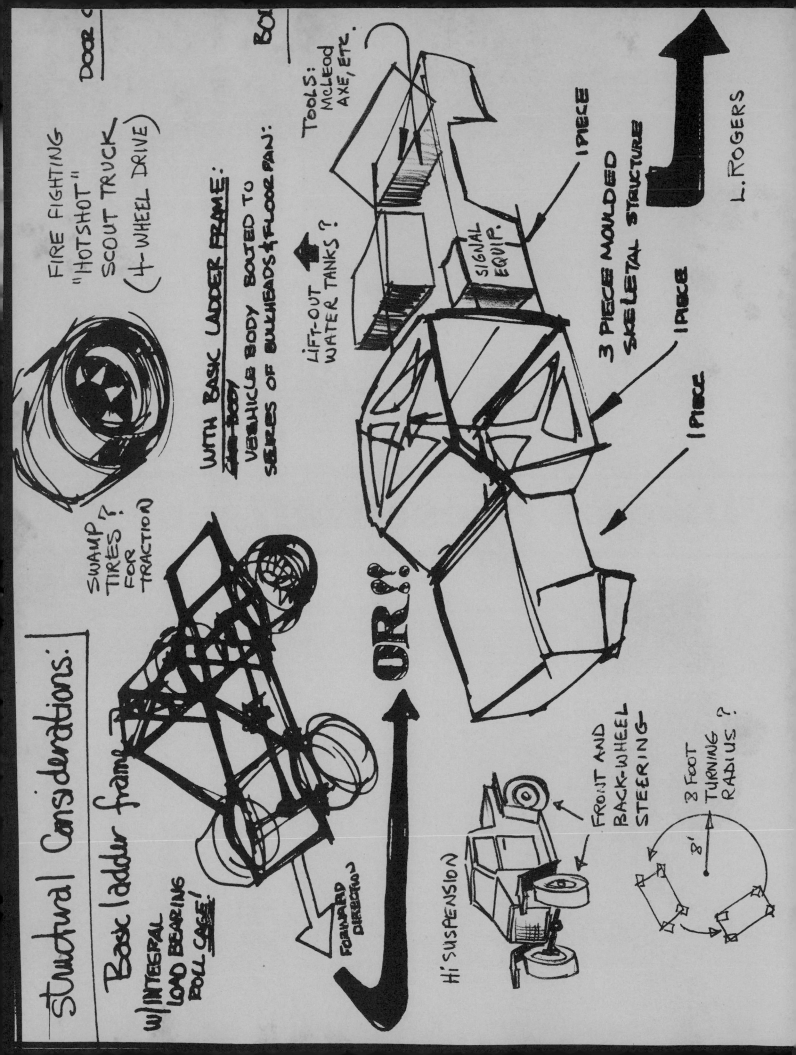

Structural Considerations:

Basic ladder frame
w/ INTEGRAL
LOAD BEARING
ROLL CAGE!

FIRE FIGHTING
"HOTSHOT"
SCOUT TRUCK
(4-WHEEL DRIVE)

SWAMP
TIRES?
FOR
TRACTION)

WITH BASIC LADDER FRAME:
Sub-Body
VEHICLE BODY BOLTED TO
SERIES OF BULKHEADS & FLOOR PAN:

Tools:
McLEOD
AXE, ETC.

LIFT-OUT
WATER TANKS?

Door

SIGNAL
EQUIP.

3 PIECE MOLDED
SKELETAL STRUCTURE

1 PIECE

1 PIECE

1 PIECE

L. ROGERS

OR!?

FORWARD
DIRECTION

HI' SUSPENSION

FRONT AND
BACK-WHEEL
STEERING-

8 FOOT
TURNING
RADIUS?

8'

SECTION 8 DRAWINGS & PRESENTATIONS

One might divide this section roughly into three parts:

1. Design Drawings

   These visualizations are made by the designer for himself or other inplant personnel to formulate the concept, establish dimensions and visualize the end product.

2. Sales Presentation

   The presentation to an outside client is usually more formal and is sometimes executed and mounted by a professional illustrator. The client then bases his decision to produce on the visualizations and related information.

3. Production Drawings

   The drawings given to a production staff are more technical and detailed. They would include the working drawings or blueprints, lists of materials, and assembly specifications.

In all three of the above cases some SCALE must be decided upon. Whereas a toy might be drawn full scale (1 inch = 1 inch), it is rather difficult to draw a railroad tank car or sailboat elevation full size. A scale drawing such as:

Half scale	1/2 inch = 1 inch	or	6	inches = 1 foot
Quarter scale	1/4 inch = 1 inch		3	inches = 1 foot
Eighth scale	1/8 inch = 1 inch		1 1/2	inches = 1 foot

allows for a more comprehendable visualization.

DESIGN DRAWINGS

 Sketches and Diagrams

The designer starts sketching and diagraming by merely EYEBALLING without scale. (See examples on previous page.)

A. Diagrams

Are rather abstract with the purpose of showing an idea or relationship or diagnosis of the problem. (Can be rough or finished.)

B. Sketches

Are ideas that start being limited. Rough illustrations. Usually involved more with perspective and a 3-dimensional, realistic appearance. Used for persuasive reasons as well as for determining directions for self. Exploded sketches would be included in this category.

EXPLODED DRAWINGS

As the young designer gets into more complex projects he will find that preliminary exploded sketches help him understand the three-dimensional aspects of his design. A first problem might be to redesign some machine tool by making some major change in idea but using the basic parts and attempting to simplify. For example, he could scrounge an old skill saw out of a junkyard or second-hand shop, dissassemble it and lay the pieces out in order on a big sheet of wrapping paper. Now instead of merely redrawing the machine, take out the electric motor and replace it with one run by butane, let's say. This saw could then be carried in the woods by a forest ranger without having to lug along an electric generator or gasoline. Safer, lighter, and simpler? Below might be the first exploded sketch. This type of redesign problem starts the student thinking in three dimensions, and he soon discovers that it isn't as difficult as it appears at first sight. From these first sketches will come the sketch models. Then parts, housing clearances, handles, controls and elevations are gradually worked out until a presentation model can be built.

Why would a forest ranger carry a skill saw? Well, then you find a junked chain saw, and you think up the major design change.

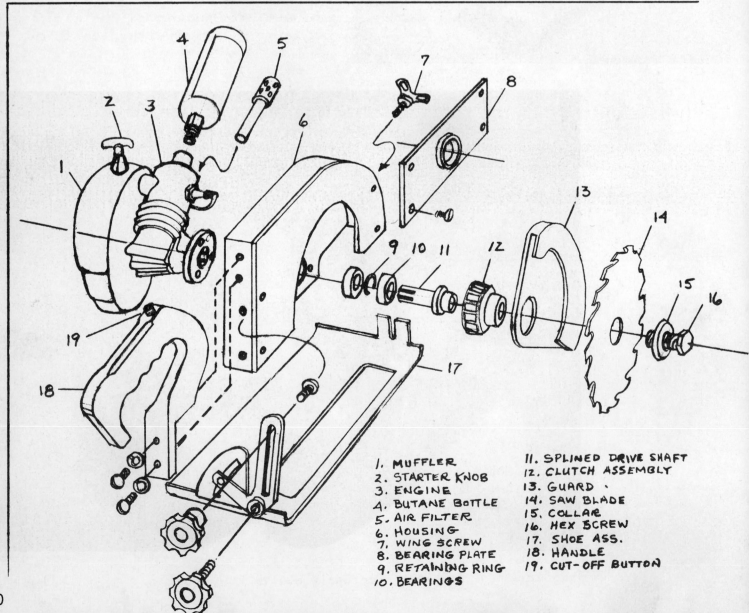

1. MUFFLER
2. STARTER KNOB
3. ENGINE
4. BUTANE BOTTLE
5. AIR FILTER
6. HOUSING
7. WING SCREW
8. BEARING PLATE
9. RETAINING RING
10. BEARINGS
11. SPLINED DRIVE SHAFT
12. CLUTCH ASSEMBLY
13. GUARD
14. SAW BLADE
15. COLLAR
16. HEX SCREW
17. SHOE ASS.
18. HANDLE
19. CUT-OFF BUTTON

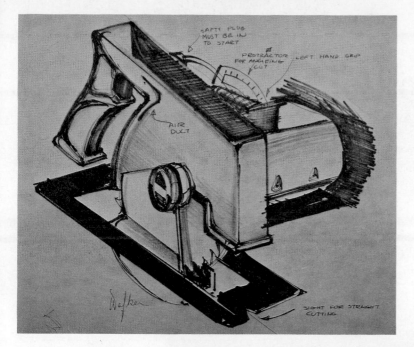

Doug Walker
Instructor: Herb Tyrnauer
California State University, Long Beach

Doug Walker   Sketches eventually evolve into model.

The pro-designer never uses rulers, ellipse guides or other templates when searching or thinking out his ideas. The straight edges and templates are used to firm up and construct his idea in proportion or function AFTER he has the concept established. Stay loose and fast while you think on paper.

Beginning designers are encouraged to improve their skills constantly in sketching and drawing. A student should be able to think with his sketches and talk with his drawings. At the same time he might do well to try quick 3-dimensional "sketches" (See the half-scale construction paper rough in Project 4, Section 3.) and perhaps one or two quick, simple modeling problems.

C. Scale Elevations

But as soon as he jells an idea, he has to draw it to scale to see what the proportions will look like, what its real dimensions will be, and whether it will work functionally as well as esthetically. The first scale drawings are usually the ELEVATIONS. ⟶ ⟶

D. Perspective Views

From the scale elevations he may draw a few PERSPECTIVE views, modify and redraw until he starts firming up the dimensions on the elevations.

E. Renderings or Illustrations

When he feels good about all the views and dimensions he may knock out a Canson paper RENDERING or other color illustration for the boss or inplant personnel.

Canson Paper & Prismacolor (Wax) Pencil

The basis for any good rendering is a good drawing. After you have drawn a perspective view of the product on bond or tracing paper, white chalk the back of the drawing, place it over a piece of dark, subdued-colored Canson paper and transfer the line drawing to the Canson paper by tracing over your drawing. Use straightedge and french curve to keep lines as precise as possible while you trace. Remove the drawing paper.

Now use a white wax pencil and your templates to firm up the drawing directly on the Canson paper. As you do this, the following technique is sometimes useful and prevents the drawing from looking too irregular or "labored." Press on the white pencil as you start the line and, as you WHIP the pencil along the guide, release the pressure and flick the point into the air. The resulting line has a needle-like, hard-edged quality that simulates reflected light along the edges of your rendering. Even though you gradually add color to the surfaces of the rendering, these white edges can often be left and add a crisp appearance to the rendering.

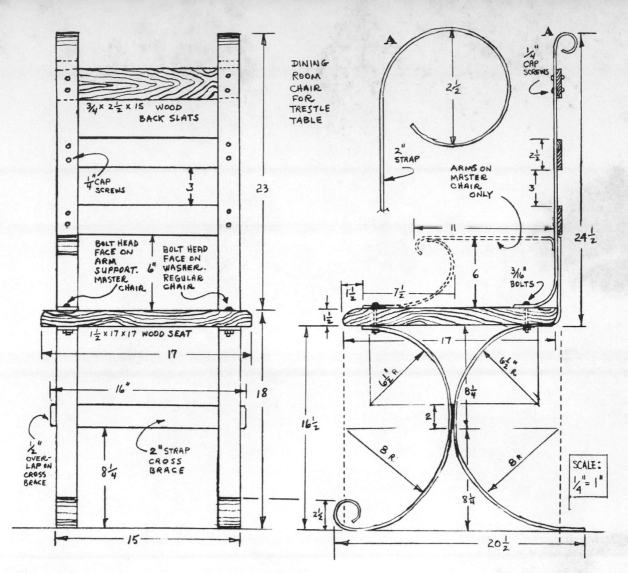

DINING
ROOM
CHAIR
FOR
TRESTLE
TABLE

$\frac{3}{4}$ x $2\frac{1}{2}$ x 15 WOOD BACK SLATS

$\frac{1}{4}$" CAP SCREWS

3

23

BOLT HEAD FACE ON ARM SUPPORT. MASTER CHAIR

6"

BOLT HEAD FACE ON WASHER. REGULAR CHAIR

$1\frac{1}{2}$ X 17 X 17 WOOD SEAT

17

16"

18

$\frac{1}{2}$" OVERLAP ON CROSS BRACE

2" STRAP CROSS BRACE

$8\frac{1}{4}$

15

$2\frac{1}{2}$

2" STRAP

ARMS ON MASTER CHAIR ONLY

$\frac{1}{4}$" CAP SCREWS

$2\frac{1}{2}$

3

24$\frac{1}{2}$

11

$\frac{3}{16}$" BOLTS

$1\frac{1}{2}$

$1\frac{1}{2}$

$7\frac{1}{2}$

6

17

$6\frac{1}{2}$" R

$6\frac{1}{2}$" R

$8\frac{1}{4}$

2

8 R

8 R

$8\frac{1}{4}$

$16\frac{1}{2}$

$2\frac{1}{2}$

$20\frac{1}{2}$

SCALE: $\frac{1}{4}$" = 1"

Elevations

The above 1/4-scale drawings were reduced to 50% of original. So the scale is now 1/8 inch = 1 inch.

Putting your elevations on tracing paper makes them easier to reproduce. If the drawings are on bond or other opaque paper they usually have to be photographed on a large "Process Camera" first to produce a negative. The negative then is laid over a light-sensitive paper, exposed to light and the image appears on the developed print or printing plate. Machines like Xerox printers, which make duplicates by bouncing light off opaque artwork, are not yet as precise as the other methods.

However, your drawings on tracing paper are like a negative in that they can be laid over any light-sensitive paper such as blue-line, brown-line or silverprint paper, which, when exposed directly to a bright light, will yield a cheap, same-size paper print, without the intervening camera process of producing a negative.

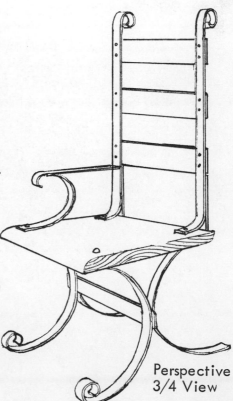

Perspective
3/4 View

123

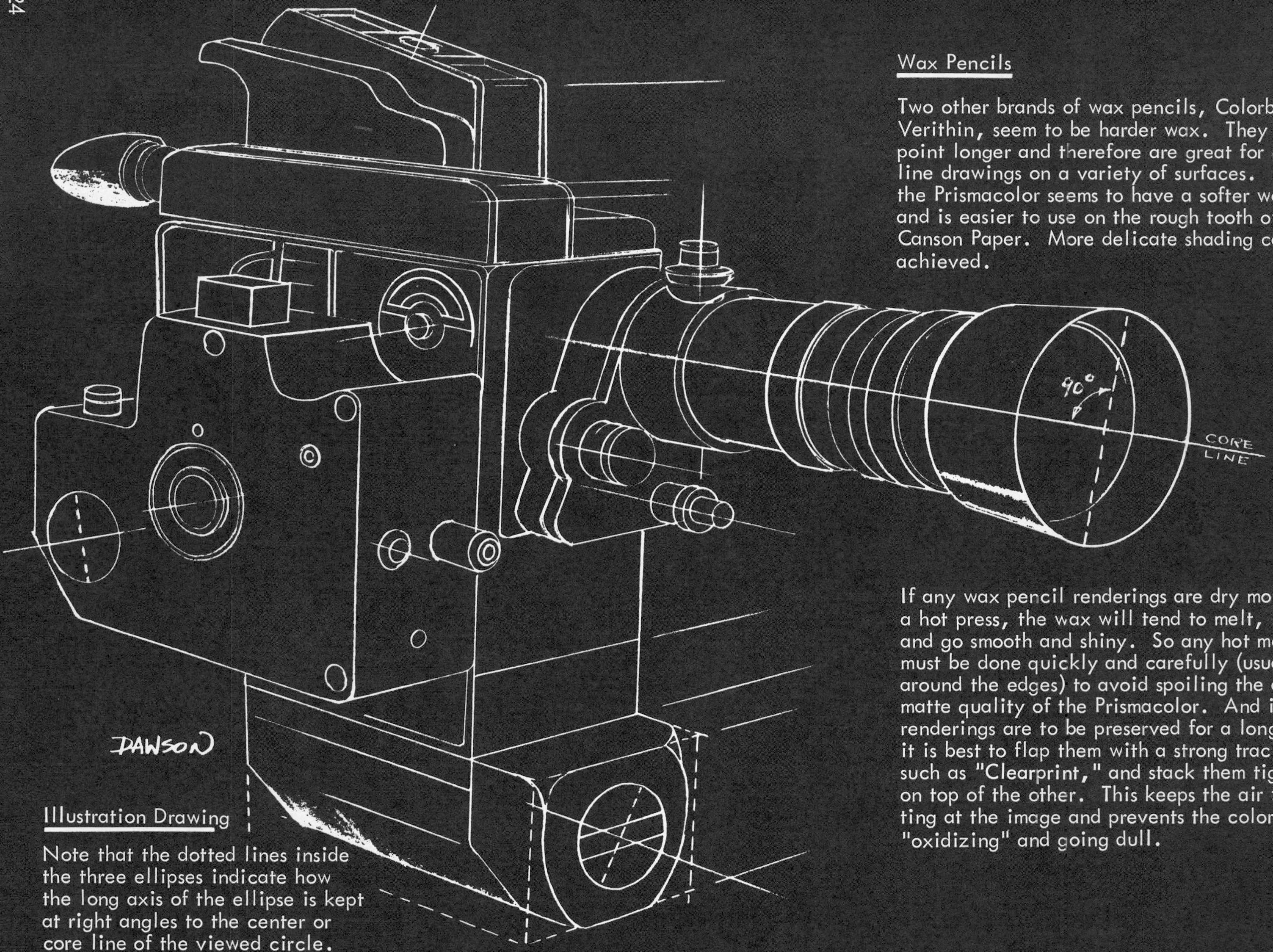

Wax Pencils

Two other brands of wax pencils, Colorbrite and Verithin, seem to be harder wax. They hold a point longer and therefore are great for colored line drawings on a variety of surfaces. However, the Prismacolor seems to have a softer wax base and is easier to use on the rough tooth of the Canson Paper. More delicate shading can be achieved.

If any wax pencil renderings are dry mounted in a hot press, the wax will tend to melt, liquify and go smooth and shiny. So any hot mounting must be done quickly and carefully (usually around the edges) to avoid spoiling the delicate matte quality of the Prismacolor. And if the renderings are to be preserved for a long period, it is best to flap them with a strong tracing vellum, such as "Clearprint," and stack them tightly one on top of the other. This keeps the air from getting at the image and prevents the color from "oxidizing" and going dull.

DAWSON

Illustration Drawing

Note that the dotted lines inside the three ellipses indicate how the long axis of the ellipse is kept at right angles to the center or core line of the viewed circle.

CORE LINE

90°

124

Instructor: Herb Tyrnauer California State University, Long Beach   Todd Dawson   Instructor: Dan Ashcraft   Pasadena City College

Doug Walker

One of the advantages of using dark Canson paper is that the color of the paper can be used to portray the majority of the surfaces. Any color added here and there to the edge of an arm on a chair or the side wall of a desk is kept at a minimum. The colored pencils should be laid on easily and carefully, generally keeping your shading lines parallel with each other and occasionally "forcing" an edge by laying in a heavy layer of color near a corner edge. This difference in value or color then helps "turn the corner" and adds to the definition of form by showing the eye that the two adjoining planes are at a definite angle to each other.

Ink & Watercolor

I will repeat again that the basis for any good illustration is a good drawing. Professionals with years of experience can visualize structure and vanishing points without doing a lot of tissue overlays, extension lines, and horizon-line planning. But the beginning designer had better take some time to check his perspective views with the instructor BEFORE he goes to his color.

After the drawing has been okayed by the instructor, cover the back with a thin coating of graphite by rubbing it with the side of a 4H pencil. (If you use softer B pencils the graphite is too thick and greasy and acts as a resist to the water color.) Now position your drawing over a piece of cold-press illustration board. Cold press has a soft vellum tooth that holds the watercolor. Hot press board is a very smooth illustration board and tends to let the watercolor run more. You may also use heavy watercolor paper or even heavy colored paper but make certain it is heavy enough not to wrinkle and buckle when the watercolor is applied later.

Tape the drawing in place over your ground and trace the lines down on the ground. You can be very exacting here by using straightedges and templates to insure every line is exactly in place. Or you can use a more free-and-easy approach by just slashing in certain key lines that you feel are necessary, such as the line of the back of the item, the base of the seat, a prominent vertical, or whatever.

Remove the drawing. Now you can copy the drawing on the board in India ink by using a Rapidograph pen or equivalent and a transparent straightedge. Or you can approach the drawing with a more flexible pen point, fine or broad, and freehand in the chair by using "looser" strokes and dabs or "deerfoot" slashes for shadow emphasis. (See examples.) A fine or broad felt-tip marker will produce a different effect.

Keep the number of colors you work with at a minimum during your first efforts. Also, select the subdued chromas rather than the brights at first. Watercolor is transparent and allows the light to go through it, hit the white reflecting surface of the illustration board and then bounce out to the eye. This gives your illustration plenty of BRILLIANCE. Yellow ochre will look like gold; a brick red which looks dull on the palette will be plenty bright when applied to the ground. Take it easy and mix your colors always with a bit of their complement, or your illustrations will look as garish and out of taste as a comic book page.

Loose Pen

Precise Pen

127

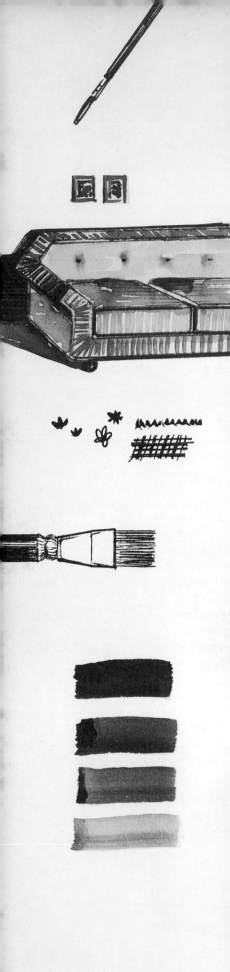

Now spray fix the ink or marker drawing with a "Workable Fix." Workable means the fix spray dries with a vellum or toothy surface that takes watercolor. (You can work over the ink drawing without fixing it, but you have to be a little more careful, as the ink will run somewhat when the water touches it.)

You can also place your color washes over the inert pencil drawing and then lay the ink lines over the washes when they are dry.

After the ink has dried, lay in a few watercolor washes over the item. It is usually to your advantage to have prepared a few drawings on scrap pieces of illustration board to try out a few strokes before you work on your main piece. A number 8, pointed sable brush gives one type of calligraphic brush mark that is often applicable to drapery designs, flowers or brightly covered furniture. It is also useful for placing small spots of color very precisely in the smaller areas. A 3/4-inch wide, flat, square watercolor brush will give a very different effect of blocks of color, which is often effective when describing appliances or machinery which are rectangular in nature.

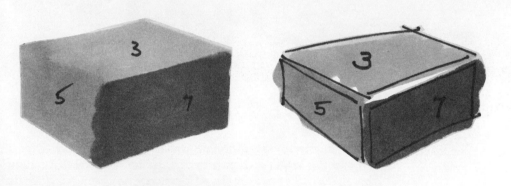

Lastly and most important of all: Watch your color VALUES as carefully as you watched your shading values in those graphite pencil renderings you did. Remember that form is achieved only through the eye seeing a CHANGE OF PLANE. Keep in mind then as you apply color that the darker colors should generally be kept on the side of the rendering away from your chosen light source, medium-value colors on another set of planes, and the lightest values reserved for the highlights or those areas facing the light. A dark blue mixed with Burnt Umber makes a good accent dark. Use this "black" to accentuate crevices, dark shadows or selected edges where emphasis is needed. The reason most artists don't use black for mixing colors is that black is a "sink." That is, it absorbs all light and makes most colors look charcoaly and dead. Real shadows reflect some light. To simulate the lighter shadows start with the hue of the material and add a complement to it. For example, violet added to the yellow wash on the side of a yellow oak cabinet, or a bit of dark blue-green added to the reddish-orange wash applied to brickwork will produce shadow-like tones with life and light-reflecting capacity. Leave black off your palette for a few years.

Felt Markers

Marks made with felt markers are pretty permanent. That is, they don't lend themselves to much manipulation like watercolor or colored pencils. When they're down, they're down. With this in mind then:

When doing realistic renderings with markers, TAKE YOUR TIME. Think a little longer before each stroke. Even try the stroke on a scrap piece of your ground before you attempt the stroke on the final comp.

We will assume you have completed your drawing and transferred it onto your ground. Instead of trying to lay in the marker swatch freehand during your first attempt, lay a transparent straightedge, like a clear plastic triangle, along one edge of the drawing, place the marker against it and pull it firmly along the guide. Pressing at the start and finish will leave a dark mark. Releasing pressure on the marker just before you finish the stroke will leave the end smoother.

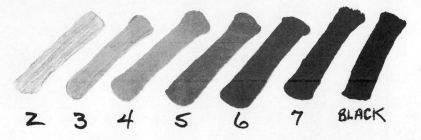

2   3   4   5   6   7   BLACK

Use ships curves for curving areas, or in some cases you may even want to cut out your own template to fit a certain configuration in your drawing to insure the felt swatches remain parallel and don't get too random or crisscrossed. Taping pennies under the template will raise it up off the paper and keep the ink from seeping back under the template. Darker swatches laid in at the sides or bottom of objects or under legs help give the impression that the item is setting solidly on the floor or against a wall.

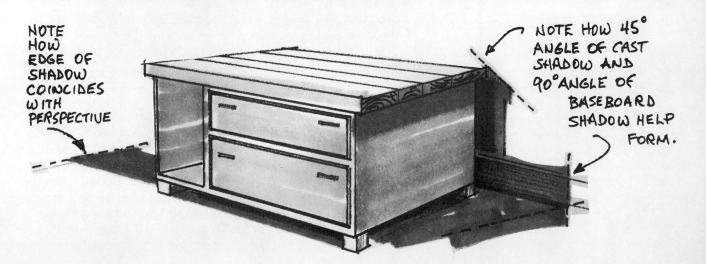

NOTE HOW EDGE OF SHADOW COINCIDES WITH PERSPECTIVE

NOTE HOW 45° ANGLE OF CAST SHADOW AND 90° ANGLE OF BASEBOARD SHADOW HELP FORM.

And check your values. Try to think in terms of 3 or 4 values only. Lay out a set of the markers that you think will give you 3 separate values. Try them on a quickly sketched box to see how they relate. Is the dark side too dark? A light-oak cabinet might seem more real if the shadows weren't as dark as the walnut chair next to it. Within each value you could have more than one marker. For the oak cabinet you might want a light yellow and a light yellow-orange for the lightest side. Perhaps two shades of yellow ochre and a raw sienna for the middle values plus a raw umber and a golden ochre for the dark sides might be blended to produce flat values over which a thin brown ink line might be added to simulate wood grain.

Fix spray will soften, darken, and blend chalks and pencils. To simulate plaids, for example, crosshatch several bright colors and then "wipe" spray (quick pass) until colors blend.

Holding triangle or straight edge at an angle with paper will give better control and prevent bleed-under.

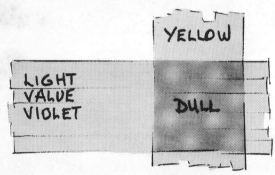

To dull a marker color (say, yellow) lay a LIGHT value of its complement down first. Then work over this light violet with the yellow marker and you will get a muted, subdued yellow. Good for shadows.

Use discretion regarding CAST shadows. The darker values on the object are one kind of shadow, but the shadow cast on the floor or wall is another kind of shadow. Shadows are absences of light, and the shadow cast by the yellowish-oak cabinet on a gray-blue wall will be gray blue, and on a pink wall will be grayed pink. The color of the object casting the shadow has little to do with the color of the cast shadow. (Only when the cabinet gets so close to the wall that the yellow light is reflected onto the wall will the local color of the object have any influence on the color of the shadow.) Now, of course, if you feel as an artist that the cast shadows look better if they have a slight yellowish tone near the oak cabinet, go ahead. But be aware that you are using the color in an esthetic way, not necessarily in a rendering way. If the cast shadow helps define the form in some way, it may be useful. But if it doesn't serve any purpose, why not leave it out? Beginners always seem to get so involved with lots of cast shadows that these gray, jigsaw backgrounds often drown out the item itself. They may help the composition in an abstract way, but don't overdo it. Simplify. Remember for the 10th time: The client wants to see the product, not the shadows.

If you have trouble making the markers work for you, try the following exercise: Select a few items from a newspaper, like a stereo set, automobile, camera, or what have you, and lay a piece of tracing paper over one of them. Then, using a set of gray markers (light to dark) work over the tracing paper, choosing the correct value marker and laying in the tones to match those of the newspaper illustration as closely as possible. When you have a marker indication completed on the tracing paper that is BETTER than the newspaper ad itself, try a few indications by choosing a set of colors whose values approximate those of the gray marker set. All of a sudden you will begin to see how the value of a color is much more important in developing a form than the hue itself is. When you are satisfied with a few of these exercises go on to the illustration of your own product.

Dipping marker tips in rubber cement thinner will soften strokes, or going over the colored area with a facial tissue moistened in thinner will lighten the value also.

Back Marking

Another technique with the felt-tip markers is using them on the back side of heavy tracing paper, frosted acetate, or clear acetate sheets. Do your perspective drawing on the tracing paper with a rather heavy line. Then turn the tracing over and apply your colors on the back side of the paper. When the colors are applied on the front side of the drawing they sometimes tend to obliterate the lines, whereas when placed on the backside they remain behind the lines, and the structure of the item remains a bit more obvious. Drawings may be done on the frosted side of acetate with the colors applied to the back, shiny side. Or you may wish to lay a piece of clear or frosted acetate over the line drawing and lay down your color right on the acetate sheet. Then when you pick it up there will be no lines at all, merely the color swatches. Over this color indication then you could come back in with an acetate ink line. (Regular India ink will not hold on acetate or other slippery surfaces.)

Instead of the markers you could paint the back of the acetate sheets with acrylic paints. This is particularly effective, as the acrylic paint spreads out into a very smooth flat finish when viewed through the acetate.

Chalks

If the square chalks (pastels) are used for presentations, the same considerations of value and shadowing should be observed. However, most chalk work should be done on bond paper ON THE BOND PAD. The resilient surface of the thick pad allows greater control and value variation. (Chalks do not work as well on illustration board.) The completed bond sheet can be dry mounted on illustration board later, or matted as is. Fixing the chalks makes them several shades darker. And don't hold the spray can too close or the fix will puddle, sog up and run. Stay at least 6 to 8 inches away. Start the spray off to one side and then "wipe" spray back and forth. Wait 10 seconds before applying second coat, etc. Several light coats are better than one heavy, soggy one.

If very exacting renderings are necessary for intricate exploded drawings, or complex projects like space capsules, the job is done by a technical illustrator. But the above techniques will at least give you a few ideas how designers present their ideas with speed and clarity.

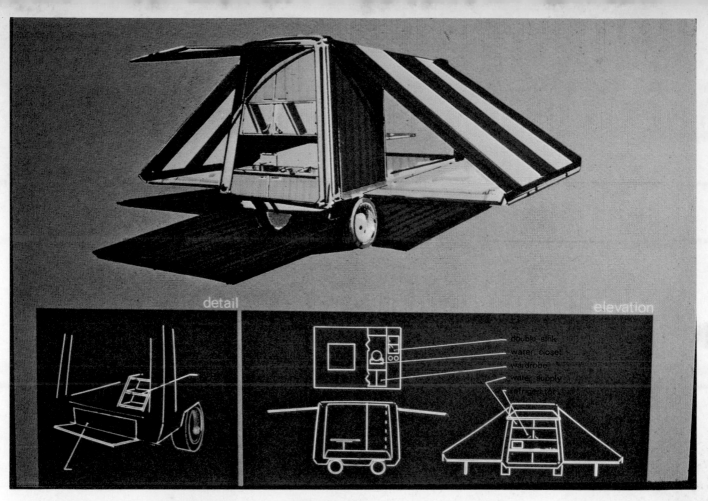

PRESENTATION BOARDS

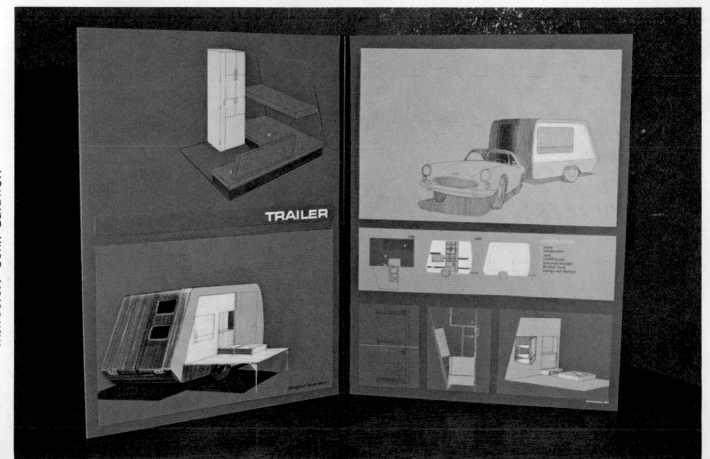

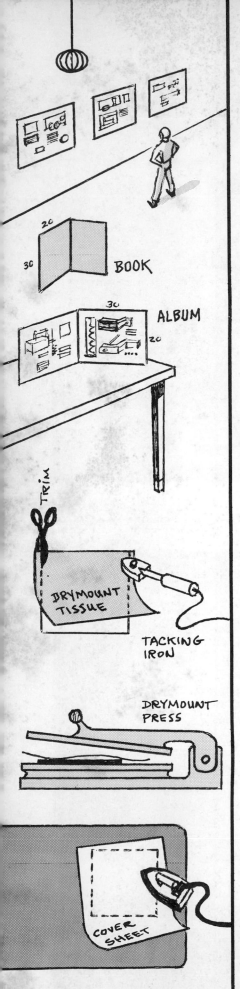

PRESENTATIONS

Presentation Boards

The presentation of a new product or the redesign of an old one is often accomplished by mounting the renderings and other pertinent information on illustration or mat boards so they can be placed on easels or table tops.

Mat boards are 40 inches by 30 or 32 inches wide. Sometimes the whole mat board is used, or the boards may be cut in half and used as two 20 x 30 boards. The two smaller boards can be taped together along one edge in a book or album form which can stand upright on a table without other support. In this smaller form they can be folded or stacked in a standard black 32 x 21 portfolio and carried easily under the arm. The large 40 x 32 boards are fine for a previously set up exhibit area where the client comes to the exhibit, but when the boards have to be transported in cars, airplanes, elevators, and set up in the typical office, the 20 x 30 size is the only way to go.

Consider the neutral toned, subdued chroma mat boards rather than the white or cream-colored pebble boards which tend to suffer from thumbprintitis. The subdued color boards, like gray-green, moss green, Gibraltar gray, yellow ochre, etc., also offer a more pleasing background and better contrast to the mounted information.

Drymounting

Drymounting tissue is a thin sheet of heat-sensitive glue. Lay your rendering or other display piece face down on a CLEAN surface and put a sheet of the drymounting tissue over the back. Tack the tissue to the back of the rendering by touching it here and there with the tip of a hot iron. Turn the piece over and trim it neatly on all four sides cutting off the excess tissue.

Place the rendering face up on your mat board in the exact place you want it, cover it with a clean sheet of tracing paper (so you can see what you are doing) and tack the rendering to the mount board by touching the two corners with the point of the tacking iron. This insures that the rendering stays put while you place it in the dry mount press.

If a dry mount press is available, place a clean sheet of tissue or bond paper over all the items that are going to be pressed down, insert the board between the two platens of the press and squeeze it together for about 30 seconds. If no dry mount press, use an ordinary launderer's iron, with the setting at synthetics, and iron right over the covering bond sheet for about 10 to 20 seconds, or until it has heated the glue and bonded the whole job down tightly. One way to tell whether the items are really bonded is to curve the board toward you and put your ear close to the board. If you hear a slight crackling noise as you flex the board it usually means you haven't heated the glue long enough and it is snapping off the paper. Just put it back in or iron it for a bit longer.

If no drymount tissue or iron, then you usually use rubber cement. Coat the back of the piece to be mounted with a thin, SMOOTH coat of rubber cement. (On thin papers do not use rubber cement that has been thinned down with an excess amount of rubber cement thinner, as it will creep through and make oily looking spots.) Now coat the surface of the mat board with a coat of thin, smooth rubber cement in the exact area you are going to place your rendering or whatever. Some excess cement can slop out over the area, as it can be picked up easily enough later. Let both coats dry for a minute or so. Register your two upper corners carefully and roll the piece down on the board. Be darn certain you have the piece in exact position BEFORE you touch the two corners down. When two pieces like this are covered with a dry coating of rubber cement, they grab immediately on contact and are the very devil to get apart. After it's down place a clean sheet of bond over it and rub it down tightly with the heel of your hand. Pick up the bond sheet and clean up any excess rubber cement around the rendering by rubbing it with a rubber cement pickup or a clean, very clean, tip of your little finger.

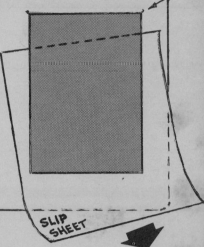

Two previously made tiny dots at each corner will help registering large sheets.

SLIP SHEET

Slip Sheeting

If your rendering is very large and hard to handle, place a clean sheet of bond paper under the rendering. Now register the two top corners of the rendering and press them in place against the mat board. Then slowly slide the clean bond sheet out from under the rendering as you gradually press the rendering down on the mat board. This is called SLIP SHEETING.

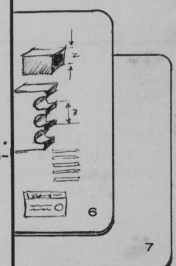

Layout

The information on a presentation board should remain rather general. It is not necessary to include all the nitty gritty details of every component. Dimensions showing overall size and a few of the more critical measurements are sufficient for most visual presentations. The materials specified can be the type of wood or metal or cloth and possibly color references, but it is not necessary to spell out the alloy composition or threads per inch, etc., on these early panels. These panels are primarily made to show the client the visual esthetics and form. In most cases details can be left until the working drawings are prepared.

The boards are often numbered so they can be placed in sequence and more easily followed while the verbal presentation is being made. Be consistent when placing these numbers, like putting them all in the lower left-hand corner.

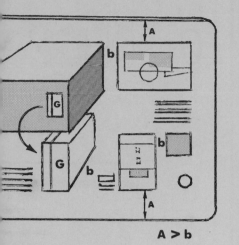

A > b

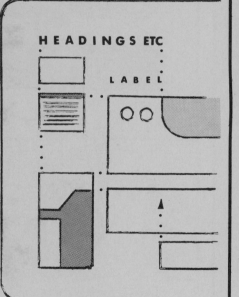

H E A D I N G S ETC

L A B E L

. . . . .
DOTS INDICATE
IMPLIED LINE
RELATIONSHIPS.

Layout (Continued)

In arranging the renderings, various views, specifications and other blocks of information, try to keep the MARGIN AROUND THE GROUP greater than the spaces between the items. This tends to give the composition unity. Otherwise the items look as if they belong to some other board, or when near the bottom of the board appear as if they were about to fall off. You should also know that your plan and elevation views should be related to each other, whenever possible, by extension lines or at least IMPLIED lines. (See Views under heading, "Production Drawings," page 156.)

Relate your images and blocks of copy by the use of this IMPLIED line technique. Line up edges, corners, centerlines or even, in some cases, edges of dark values with other similar areas. This helps keep your composition unified and prevents it from becoming chaotic and random. Simplicity is the keynote to understanding. Avoid diagonals and tricky angles for either images or type. The horizontal-vertical block type of arrangement is usually the most comprehendable. A human being's eyes shift very easily from left to right and up and down, but it is darn hard for them to sweep from upper left to lower right or other diagonal movements without making a special effort. Your head is in the same boat. Neck muscles will swing it easily form left to right and up & down, but try to make them swing your head from upper left to lower right. Surprising what an effort it takes . . . . right? Illustrators use strong diagonals for setting the mood in action pictures such as "Man overboard!" in a sailboating story or an F-105 doing a wingover, but that is the LAST effect you want. Keep it simple. Don't strive for originality and "creativeness" in your presentation backgrounds. Instead, strive for creativeness in your product, and let the "frame" be subdued and unnoticed.

Identify the pictures and drawings. You, the designer, are so familiar with all the pieces, you are apt to forget that this may be the first time the client has seen the product. Also unless the views are VERY obvious, they should be labeled also. Any detail drawings should be keyed into the larger drawing by capital letters or some other consistent key to help the viewer understand what he is looking at. The best test of a good presentation board is to show it to someone who has never seen it before and ask him if he can understand it WITHOUT YOU SAYING A WORD. The young artist is often very surprised when classmates say to him, "What the hell is that thing there?" and point to an unlabeled line drawing done in oriental perspective, with a touch of orthographic projection thrown in for good measure. It may be very obvious to you what you're drawing, but you must make it obvious to the viewer or client as well. Learn to ask your classmates if they can "read" it BEFORE you glue it down on the nice big beautiful mat board you paid $1.50 for.

Artist Aids

The black or white transfer lettering may be used very effectively when burnished off on the darker matboards for major or minor heads and captions. The rolls of adhesive-backed tape on the market are especially helpful for making borders or call-out lines. Use discretion, however. Don't start making Grecian hem stitching in the corners and "rat mazes" just because it's so much fun to run those little lines all over the place. "Ties everything together, you know." When you do use it, tack one end in place, pull the tape taut over the line you want and THEN press the far end down on the board. Now proceed to the center and tap the tape gently out from the middle until it is set. Put a clean sheet of paper over it and burnish it down tight. If you try to press it down as you go you will probably find that it wobbles from side to side and ends up looking like the stripe in the middle of the freeway ～～～～ just before you come to a stop sign. These rolls come in widths from 1/64 of an inch on up.

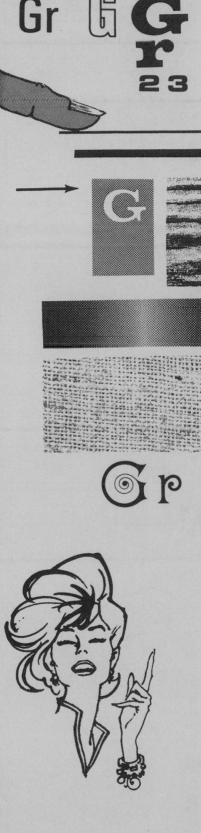

Besides the letters and lines there are numerous wax-backed adhesive carrier sheets which provide the designer with a variety of textures, patterns and symbols. These sheets cost about $1.00 to $5.00, but often save much more than their cost when balanced against an artist or draftsman time. They are available in black and white and colors in some cases. Whole sheets of transparent color can be used to emphasize important areas or provide overlays for bar graphs or technical drawings that are projected through an overhead viewer. Below are a few samples of symbols and textures. ⟶

CAUTION: Remember that many of these aids are wax backed. If pressed in a hot drymount press or ironed on they will often melt and run or twist out of position. These should generally be applied AFTER any drymounting has been done.

Slide Talks

When completed, the presentation boards can be copied on film and made into colored transparencies (slides). Slides are fine if the room can be darkened, the extension cords are long enough, the slide projector works without sticking, and someone remembers to bring the large screen. You are assured of a captive audience, in that they all have to look at the same thing you are talking about. However, there is one thing I have noticed after seeing slide presentations for about twenty years. Ninety percent of designers drag the presentation out much too long. What was intended to be a 5-minute review, with a half hour left for discussion, usually ends up a forty-minute lecture, with some slides remaining on the screen for four or five minutes until heads begin to nod, while the designer expounds and expounds and drags every last iota of incidental information from his inexhaustible wells of esthetics, psychology and philosophy.

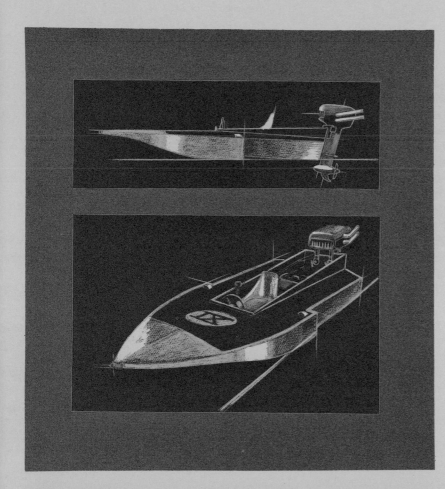

ENGINEERING CONSIDERATIONS

1. Heavy construction stops at this point with lightweight nose providing extra length to conform to California power boat minimum length of 12 feet.

2. Pine frames (3/4 x 3 in. glued & screwed together with No. 6 round head screws) are approximately 12 in. apart & support the 1/4 in. marine plywood bottom and 3/16 in. plywood deck. One inch oval head cadmium plated screws are placed 6 inches apart with waterproof resin glue in all seams, stringers, and frame contacts. Keel is 3/4 x 3 in. mahogany. Stringers are 1 x 1 pine.

3. Step (4 in. high) tends to lower water friction by keeping center of hull above water at planing speeds. Hull rides on 18 inches of stern and about 12 inches of step, so only 1/4 of the bottom is touching water at high speed.

4. Vents on each side of cockpit funnel air under step to prevent any vaccuum effect. Same air also clears bilge of gas fumes.

5. Weight distribution: Engine is fixed. Fuel tanks and driver's seat can be moved for best balance. In general, rough water requires more weight forward to prevent front end from bouncing or " porpoising " too much.

6. Minimum engine about 25 hp with dual carbs & stacks drives this hydro at 38 mph tops in glass water. A rough rule of thumb gives a speed of 1 mph for each outboard horsepower, so a 100 hp engine should better 100 mph on this hull. (If you have the nerve... eh eh.)

7. Transom is oak or mahogany 1.5 in. thick for strength at motor mount. Coaming 1/4 x 12 in. ties in to cockpit as angle brace to support transom and reduce excessive flexing of hull. Transom is angled about 16 degrees with water line which allows propellor thrust to be adjusted parallel with water surface or slightly upward when planing.

8. Cavitation plates are stainless steel tabs with turnbuckle flex adjustment which act as fixed elevators. These are helpful to control or dampen excessive porpoising.

9. Twelve feet is minimum length for power boats in California waters. Check your department of harbors before you build.

10. A wide beam is somewhat safer on rough water at high speed.

11. Vee bottom is preferred over flat bottoms in recent years. Although keel rides deeper in water, the V gives better dihedral stability and more control on turns.

12. Interior needs 4 coats of Spar varnish. Spar is better on decks also, as a resin finish tends to be too brittle and may crack under flexing stress.

13. Fibre glass strips over corners and joints saturated with an epoxy resin makes for strength and protection at seams. Three to four coats of copper bottom paint on hull or a mixture of powdered graphite and Spar varnish will give a beautiful almost frictionless surface.

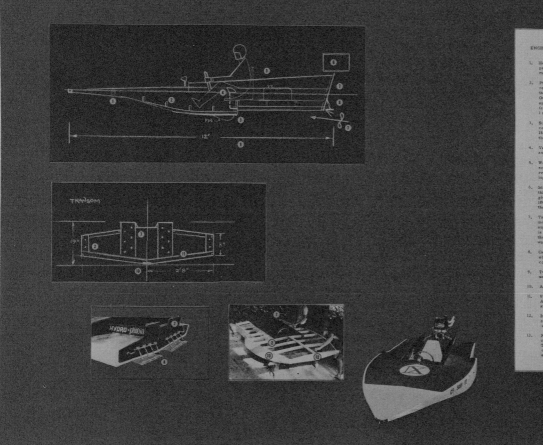

138

(See back cover for prototype.)

For Pete's sake, remember you are usually dealing with men whose time is probably MORE valuable than yours. They are well aware of most of the dead horses you are beating. Give them credit for being at least as intelligent as you. Keep the slides moving. Rather have someone say occasionally, "Hey, could you back up to number 4? I'd like to see the strut on the undercarriage before you go on." than have everyone bored to tears. Don't take for granted that you know how to show slides. Try a dry run first and have someone time you. You may be surprised what you thought was going to take 10 minutes took close to thirty. The designers who can stick to a strict time schedule will find they have a refreshed, enthusiastic group of eager people ready to ask their OWN questions concerning the design, instead of arm-stretching, yawning, exhausted bodies wondering, "When do we eat?" and thinking, "My Gawd, can that guy blab!"

There is now available a projection screen called EKTALITE by Kodak that is 6 times brighter than the average screen. It is slightly concave and has such a bright image that it can be used in an undarkened room. This is particularly helpful when making presentations in bright offices, factories or outdoor situations. Also people can take notes and facial expressions can be seen.

Board Talks

Using the boards themselves instead of slides has advantages. For example, you can see facial expressions and there tends to be more give and take between audience and designer. Clients tend to get up and move around to see details on the boards, etc. But again, remember not to talk too much. Make your presentation concise, sharp, and to the point with a bit of humor. Keep relaxed. Remember that 99% of the time there will be criticisms of this and that. Any time you ask for criticism you'll get it. And presentations are, in a sense, asking for it. Keep in the back of your mind that you and your client are a TEAM. Actually you are both after the same things: a good product, a good market and a decent profit. Listen to criticisms. Be an absorber. Stupid or unreal criticisms will be thought stupid and unreal by others in the group also. You do not always have to refute or resolve every doubt thrown at your design a split second after it is given. In a sense you are like a good chairman who goes into a meeting WITHOUT preconceived ideas of what he is going to make the group do. Go into it to find out what opinion and advice can help make a better product. Try to get EVERYONE to make some comment. And if it is a complicated product be sure to take notes, so you can sum up the total conclusion at the end and be able to use the notes as a structure when you arrange the next presentation with revised model or prototype two. Good luck.

Flip Charts

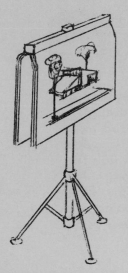

Sometimes flip charts are more practical for presentations. The big charts (4 feet x 3 feet) are often used for large audiences, such as school board meetings or other community get-togethers. The preparation of these is similar to that of the presentation boards, but the information is put on strong, flexible paper sheets which are then clamped together on one edge.   Information can be placed on both sides of the paper if placed in reversed sequence. Say you had 12 charts to show. Place the information for chart 7 on back of chart 6, chart 8 on back of chart 5, etc. Then after chart 6 has been shown turn the rack around, flip chart 7 over and work back.

However, most of the time you will be showing your ideas to smaller groups - perhaps only 3 to 6 people. Since the introduction of "Instant" printing shops, many PR men and other promotion experts are using 8 1/2 x 11 flip charts made up in 3-ring or plastic spine binders. (See illus.)

In lots of 50 pages, for example, the price for one 8 1/2 x 11 sheet printed 2 sides is about 10 cents at present. So a binder with 20 pages of information plus several photos slipped into glassine holders might cost less than $5.00 a piece. All the Instant Printer needs is an 8 1/2 x 11, black on white image, which he shoots same-size, makes his plate and runs off your 50 copies within minutes. Check with him first for complex copy or graphs.

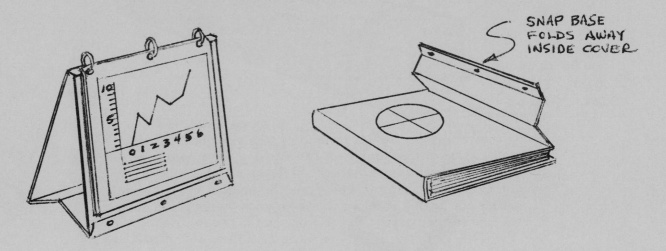

SNAP BASE
FOLDS AWAY
INSIDE COVER

During a presentation then, one of these binders can be placed upright on the conference table and flipped while the designer talks. And the neat thing about these small, inexpensive binders is that, when the conference is ended, the designer can give one to EACH person at the table without killing his budget. These are particularly useful when marketing statistics, time schedules, and production graphs are part of the presentation proposal. The conference gives each person the overall view. But then each rep or manager can go through the binder himself and talk it over in his own office without having had to take down a lot of notes or wonder what that guy really said last Friday. Also it makes your design group look efficient, and efficiency is rated right along with price and esthetics in bidding competitions.

EXHIBIT STRUCTURES

Design offices are often asked to design exhibit areas for client's presentations. There are a variety of prefabricated fittings and structural materials on the market which lend themselves to reducing the amount of labor involved. A typical one is outlined and illustrated below:

The Apton Speed Frames are 1-inch square hollow conduit with 1-inch cubes used as corner fittings. The conduit and fittings are available in different sizes and in all colors of enamels, anodizing, and chrome. The only tools necessary are a soft mallet and a hack saw. Accessory equipment includes slotted conduit for shelf supports, caps, slider feet, casters, curved sections, and snap-fit glazing channels which fit over the conduit and provide slots to hold panels of glass, plastic or plywood. One distributor is Dexion, Inc., 111 North Central Ave., Hartsdale, NY, 10530. Fittings are designed to build temporary structures for quick disassembly, temporary structures with more rigidity where quick disassembly is not important, as well as a permanent type of structure.

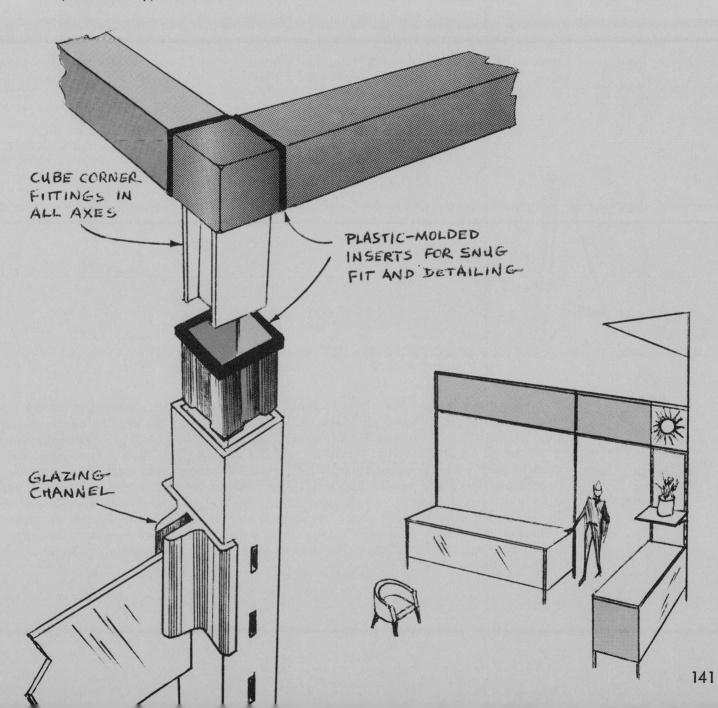

CUBE CORNER FITTINGS IN ALL AXES

PLASTIC-MOLDED INSERTS FOR SNUG FIT AND DETAILING

GLAZING CHANNEL

ILLUMINATION

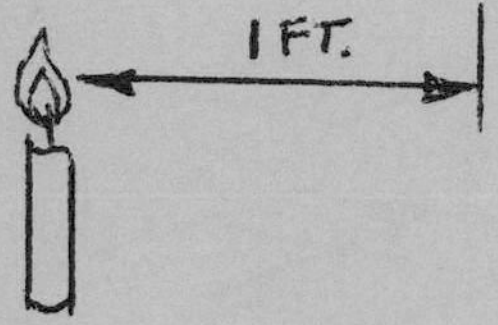

Lighting is a unique problem usually left to an architect or lighting engineer. However, there are a few suggestions or rough rules-of-thumb below that may help the young designer understand the problem or at least help him ask intelligent questions regarding exhibit lighting, etc.

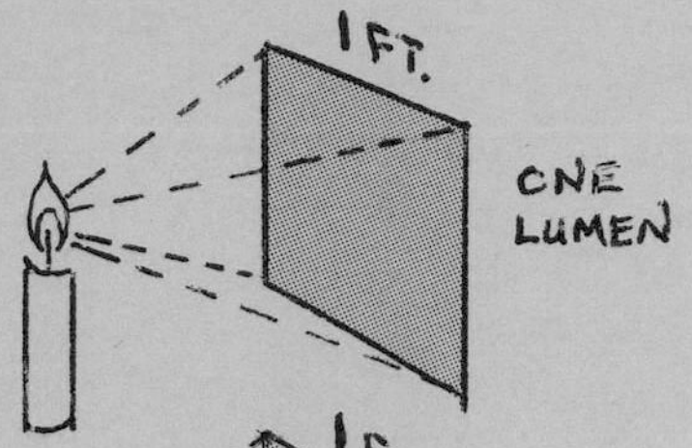

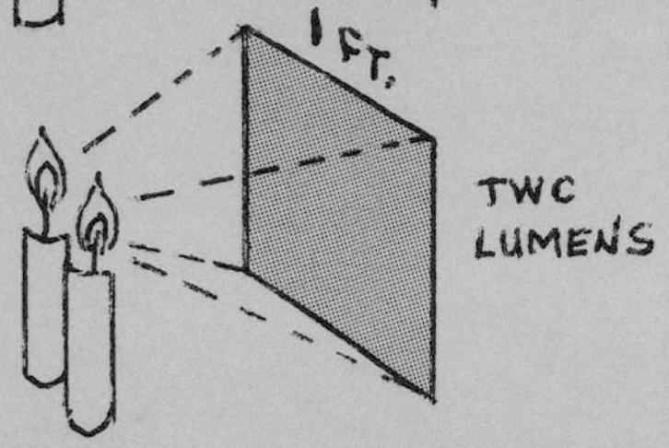

Candlepower	–	The light-giving intensity from a candle made to certain specifications
Footcandle	–	Intensity of light on a surface 1 foot away from the flame
Lumen	–	Light that falls on one square foot of surface on which the intensity of illumination is one footcandle

Recommended general footcandle illumination for a variety of activity areas have been compiled in charts. A few are indicated below. (Note that the recommended illumination increases as precision of work increases.)

Lighting is generally averaged at a 30-inch level above the floor, where most table tops, counters, work or study areas are located.

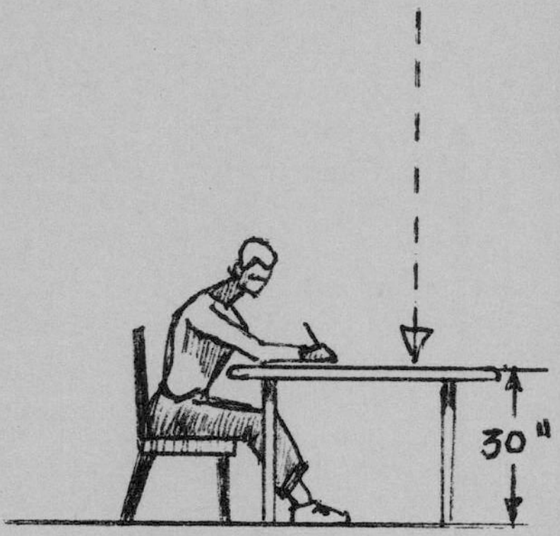

Recommended Footcandles (per square foot at 30 inches above floor)

Auditoriums	10	Printing Plant	50
Locker Rooms	10	Auto Repair Shop	50
Lobbies	20	Drafting Room	50
Living Rooms	30		
Classrooms	30	Science Laboratory	55
Gymnasium	30	Hospital Operating	55
Hockey Rink	30		

The following chart gives the rough rating in lumens of a few typical incandescent (hot wire) or fluorescent (excited gas) lights.

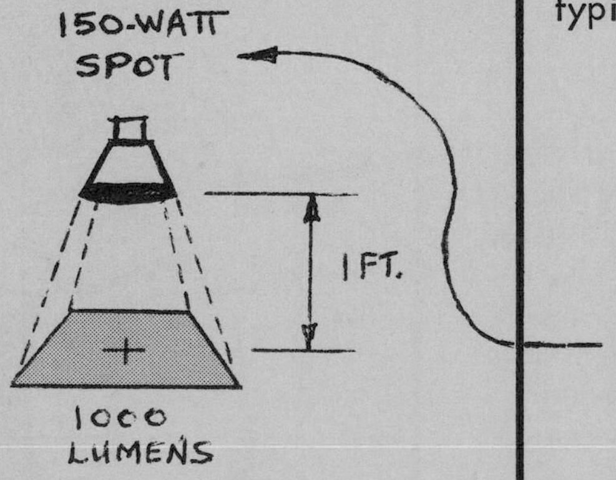

Incandescent		Life Hrs.	Lumens *
100 watt	Bulbs	750	1600
150	Bulbs	750	2600
200	Bulbs	750	3600
150 watt	Spot	1000	1000
300	Spot	1000	1400
150	Flood	1000	1200
300	Flood	1000	1600
Fluorescent			
40 watt	Tube	2500	2300
100	Tube	3000	4000

* Lumen Rating means the number of candles illuminating a one-square-foot area one foot away from the rated light. (See example of 150 watt Spot.)

As general lighting fixtures are usually hung 10 or 12 feet distant from exhibit tables, there are fewer lumens reaching the display. The intensity decreases as the square of the distance. (See diagram →→→ of the decreasing lumens from the 150-watt Spot at increasing distances.)

1000 L
1 ft.
250 L
111 L
62.5 L
40 L
6 ft.

To determine the lumens reaching the display, measure the Distance in feet (D) from the light to the display; square it (D^2) and divide the answer into the Lumen Rating (R) of the light.

$$\frac{R}{D^2} = \text{Lumens reaching display}$$

Our 150-watt Spot at 8 feet above table:

$$\frac{1000}{8^2} = \frac{1000}{64} = 15 \text{ lumens}$$

But say you decide you need 30 lumens for your exhibit (something equivalent to the recommended footcandles for classrooms) then you will need TWO 150-watt Spots hung 8 feet above the display.

This formula applies only to the area immediately under the fixture or light. Displays 10 or 15 feet away obviously won't be getting enough illumination from this one source.

Another rough method for large areas where fixtures are hung between 8 to 15 feet overhead is:

Total lumens needed for a room area are found by multiplying the square feet in floor area by the number of footcandles recommended for that activity area:

Total lumens needed = square feet × recommended footcandles

Now, because ceilings become dirty, walls tend to become grimy and darker with age, bulbs or tubes get dusty and dirty, light shades or frosted glass covers absorb some light, air may be smoky, objects cast shadows, dark furniture or draperies absorb light, etc., there are other charts that give factors to correct for these light "sinks." These Room Indexes, Maintenance Factors, etc., given in decimals all get rather complicated. However, we can apply a very rough rule-of-thumb and merely multiply our answer in the formula above by 1.2. The formula then for, let's say, an 18 × 24 exhibit area with a 12-foot-high ceiling would be :

Lumens needed = (18 × 24) × 30 × 1.2 = 15,552

You decide to use 100-watt fluorescent tubes rated in the chart at 4000 lumens. Divide 15,552 by 4000. The answer is 3.9 or, say, 4 tubes. Therefore your recommendation would be four of these tube fixtures spaced equally over the floor area.

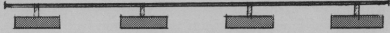

Remember that the above rough calculation would probably cause a lighting engineer to shudder. As ceiling heights change, balconies cut off rays, or narrow corridors and stairwells become part of the design problem, the mathematics involved become much more sophisticated.

Spotlighting

Besides general illumination there is spotlighting for specific work or exhibit areas. This is best arranged by trying various locations, angles, and wattage. Objects lit from the front look flat, because the light bounces off the object away from the observer on the sides and right at him from the front. Thus there is very little difference in shading and the eye cannot pick up any form. When a single spot is used from the side it shows more form but causes dark shadows on the far side. Again form is lost because there is very little light coming from the dark shadows, and the eye can't see any form there either. A smaller light or reflector on the shadow side will fill in the dark shadow areas with some subordinate light and give better form to the display.

* Advanced Students: Use fill light when fotografing your own models.

Fill

Spots particularly should be checked to make certain they are not "spilling" light into unwanted areas or into the audience's eyes. Colored Gels or Cells (Gelatins or Cellophane) over the spots should be used with discretion. Avoid bright colors like dark purples, blues and reds. You will ordinarily discover through long experience that mild, warm colors like pinks or light red-oranges are more appealing to most people than the harsher cold colors. Also, don't make the beginner's mistake of getting such a fantastic, complicated lighting system that every customer goes away commenting on the neat lighting display and forgetting what designs were exhibited. Use lighting to quietly enhance the product's characteristics, NOT to put on a light show. The best lighting is the type that is not noticed.

Model Making

One could write a whole book on model making and still not cover the necessary instructions for the very first thing a student had to build. Model making, for the student designer, requires first of all, a large dose of ingenuity. Cutting flat cardboard into patterns that eventually evolve into 3-dimensional forms, locating a contact paper that simulates the right texture or pattern, scrounging an old knob and repainting it to look like chrome are all part of the problem. In professional model shops experienced craftsman have at their disposal lathes, drill presses, table saws, micrometers, threading machines, etc. The student, however, has to rely on his originality and constant observation of trash cans, junkyards, and hardware stores to "produce" his baby.

Designers do not usually become model makers. However, the making of models for the student designer is more than the production of his design. It is often one of his first experiences in manipulating metal, cardboard, adhesives, glass, or what-have-you. It gives him an appreciation of what professional model makers go through. It give him a better feeling for radiuses, joints, and finishes. And when the model is complete, he is sometimes surprised at the number of modifications he has thought of during its construction.

Scoring Jig

When using cardboard to make model forms there is a simple jig which can be used to insure smooth-looking corner edges. Select two pieces of heavy cardboard about 15 inches long and 3 inches wide. Tape them down parallel to each other about 1/8th inch apart on another scrap of cardboard. (See diagram below.) Lay your cardboard pattern over the jig with a corner or score line over the groove in the jig. Use a hammer and drive a metal strip (Approx. 16 inches long, by one-inch wide, by 1/16-inch thick) down into the groove so that the pattern cardboard is crimped into the groove. Remove the pattern and fold. You will find that the cardboard will bend easily without tearing or wrinkling and provide a very professional, smooth look to the corners of the model.

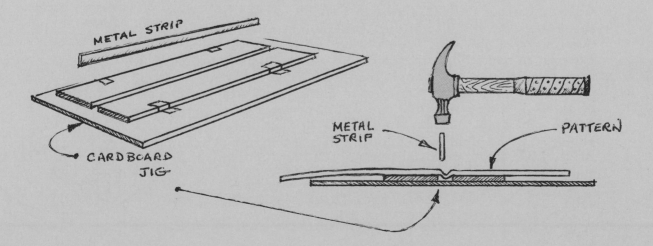

METAL STRIP

CARDBOARD JIG

METAL STRIP

PATTERN

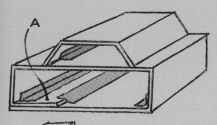

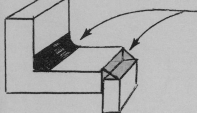

Masking or mastic cloth tape laid along the inside edge of the bend will provide stronger support for the longer edges.

When you cut your first patterns leave exceptionally large, wide tabs. Then when you fold and join edges your adhesive will have plenty of area to hold to. Narrow, dinky tabs fail to hold and as they pull out make the model look weak and distorted. (See A.)

Plastic wood, wood dough, body putty (from car repair shops), and Spackle (a white plaster-like material) are all useful in filling cracks or making fillets between surfaces. When hard they can be sanded and finished like wood.

Use the spray fixes, lacquers, or acrylic clear sprays to cover your model material before you paint. In some cases it is good to spray the model before you use the fillers mentioned above to prevent any surplus liquid media from being absorbed into the cardboard or wood and leaving a stain.

Plasticene is a greasy modeling clay that can be used to make pieces in preparation for casting. It can also be used to model the entire product. It is firm at room temperature but becomes softer and workable from hand heat. After the modeled piece has become firm it should be sprayed with a shellac or acrylic fix to contain the grease. The piece can now be painted without fear of the grease from the clay seeping through the finish.

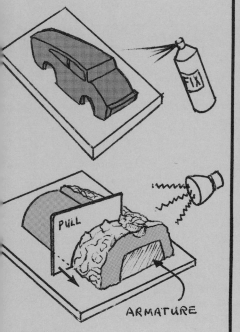

PULL

ARMATURE

TEMPLATE

Detroit Clay is a similar modeling clay. However it remains harder than the plasticene and must be heated with heat or infra-red lamps to make it soft enough to work. Templates cut from thin metal sometimes help the modeling process, particularly in producing smooth cylindrical surfaces.

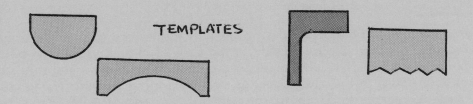

TEMPLATES

Contact papers are available in many different simulated woods, metals, and other materials. These papers have an adhesive backing and can be applied very easily to the finished form. Beware of putting the contact paper over the flat pattern before it is folded, because it sometimes tears at the corners as the pattern is folded. In some cases it is better to cut the contact paper to fit INSIDE the folded edges. Even though the corner edges are then left uncovered, the thin white line around the edges of the completed model does not seem to bother the visual appearance. On the other hand, if you try to cover all the sharp corners with little bits of tape or paper and paint, you may find the result looks tattered and rougher than if you had left it alone. Ragged edges and corners are usually noticed before anything else, because they form the outline or significant silhouette shape of the entire unit. This is one of the first things an eye scans when it looks at a product.

Styrofoam Cutter

Dense Styrofoam is a lightweight, rigid material that can be used to make a variety of forms. It is easy to cut with the hot-wire jig sketched below and is used in many cases to make up the structural parts or armature for a form that can later be covered with other materials. It has the advantage of being easy to use and fast to produce. It can be coated with latex paint and painted like any other material. Enamels will eat into foam.

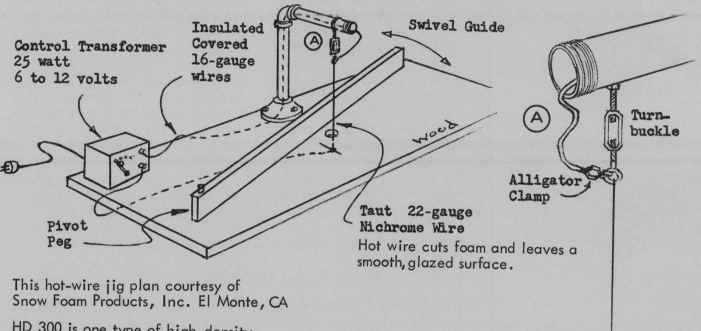

Control Transformer
25 watt
6 to 12 volts

Insulated Covered 16-gauge wires

Swivel Guide

Ⓐ

Pivot Peg

wood

Ⓐ

Turn-buckle

Alligator Clamp

Taut 22-gauge Nichrome Wire

Hot wire cuts foam and leaves a smooth, glazed surface.

This hot-wire jig plan courtesy of Snow Foam Products, Inc. El Monte, CA

HD 300 is one type of high-density styrofoam

TOOLS

A band saw and a floor-model, power belt sander are probably the minimum power tools for a model shop. Check on the ease of changing sanding belts as well as noise. Some of the belt sanders are excessively noisy. And this will drive you out of your gourd in the classroom, as most beginning students try to use the sander as a saw to remove large chunks of material, instead of using it as a finishing device only.

C-Clamps of all sizes are particularly handy in a room where there is very little equipment. They serve as vises (to table tops) or clamps to hold pieces together while the glue is drying, etc.

Hypodermic needles filled with plastic cement make an excellent "brush." The two pieces of acrylic are clamped together as desired (see sketch) and then a thin line of cement is run along the crack between the two pieces with the tip of the hypodermic needle. The cement flows in between the two pieces and provides a good clean bond with very little excess glue problem.

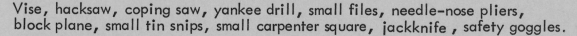

Vise, hacksaw, coping saw, yankee drill, small files, needle-nose pliers, block plane, small tin snips, small carpenter square, jackknife, safety goggles.

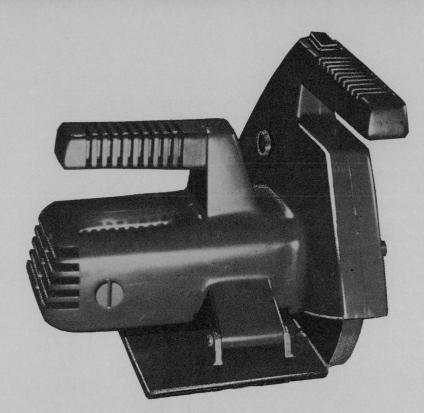

Doug Walker

Instructor: Herb Tyrnauer
California State University,
Long Beach

Mockup Model of Skill Saw

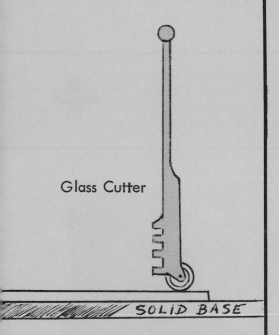

Glass Cutter

SOLID BASE

Types of Glass

Number 1
Number 2
Crystal
Plate
Tempered
Laminated (Safety)

Machining Plastics

Leave paper on sheets when cutting (can be removed later).
Lubricating the table saw blades with a medium-weight grease
or wax or parafin will reduce chipping.

    Sabre saws   Use 14 teeth per inch minimum
    Band saws   Use 10 teeth per inch minimum
    Table saws   Use 6 teeth per inch minimum

Sanding   To clean edges use 60–80 grit paper first, then
                150 grit, or polish.

Drilling   Need sharp drill, slow speed and minimum pressure.

Glass-Cutting Procedure

Wipe glass clean. Place glass on soft but firm level surface. In-
door-Outdoor carpeting tacked down over a flat, rigid table top
makes an excellent working surface. Press straightedge firmly;
hold cutter upright between thumb and forefinger and run the
wheel along straightedge with firm pressure. START cut 1/16
from edge to avoid chipping. But allow wheel to drop off edge
of pane at finish of cut. Wheel may be dipped in light oil for
critical cuts. Break short pieces between thumb and fingers of
both hands. Longer pieces can be broken apart by placing a
long piece of wood under the cut and applying even pressure on
one side. Notches on the handle can be used to snap off narrow
strips.

Old tin cans are a good source of model material. The metal can be flattened, cut with a tin snips, and folded into many forms.

Quick-hardening plastic from hobby stores comes in small cans with a hardener. If you want several similar parts, you can make several identical molds, mix some of the plastic with a few drops of the hardener and fill the molds. Overnight is usually the prescribed setting-up time. A similar material with a trade name of Sculp-Metal needs no hardener. It appears to be a fast setting resin with aluminum powder or filings mixed into it. When it sets up it looks like aluminum and is as hard as metal.

Welding rods come in a variety of diameters and colors. These can be used in many ways, especially on the scaled-down models, to simulate trim, handles, tubing, etc.

Balsa wood sheets and sticks plus small doweling are sold by most hobby stores. These are easy to cut and work with. Balsa works well with the quick-drying model airplane cement.

Liquid Solder is a quick-drying varnish-like glue that dries with a dull gray finish. Useful for light wire or metal joints.

Barge Cement holds leather.

FOME-COR is a Monsanto product available thru most paper companies. It is a sandwich laminate of Polystyrene foam about 1/5 the weight of cardboard. Comes 1/8 to 1/2-inch thick. Easy to cut, score, and emboss. Accepts all media.

 An unusual but important first-aid tip: If a nail or splinter ever penetrates the eyeball, do NOT pull it out. If you do, the liquid inside the eyeball may run out. Keep patient quiet, on his back, until doctor can get there. Then in surgery the doctor can usually save the eyesight.

There are two reference books regarding important measurements of the human body. Related information on pedals, manual controls, illumination, and auditory or visual displays are included:

"The Measure of Man" (Human Factors in Design) Henry Dreyfuss
Whitney Library of Design
18 E. 50th Street
New York 22, NY

"Human Factors Engineering Handbook"
TAD Products Corporation
PO Box 25
Beverly, Mass. 91915

Mockup Opaque Projector

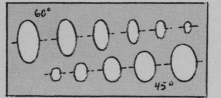

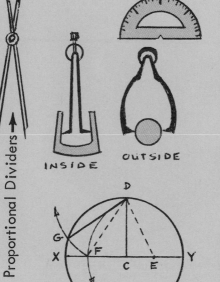

INSIDE   OUTSIDE

PRODUCTION DRAWINGS

It is assumed that the student in product design is also taking courses in freehand drawing and perspective and rendering. Many students have not had any drafting classes and are at a definite disadvantage when making up details and working drawings. Thus a few mechanical drafting techniques are given below plus a few suggestions on drafting tools, standard dimensioning practices and blueprint reading to assist him in making his detail drawings on the tracing paper as clear as possible.

Tools

Engineer's Scale

There are a variety of these three-edged rulers, each with 6 scales dividing inches into 10ths, 12ths, 16ths, etc., so that the designer can figure dimensions quickly on scale drawings.

French Curve

A plastic sheet with a variety of arcs cut into its configuration. Useful for drawing accurate short curves.

Ships Curve

Similar to the French Curve but having only two or three long, sweeping curves for longer curves.

Ellipse Templates

Most circles are viewed as ellipses. The template is identified by the particular angle a circle is viewed from and contains a variety of widths.　Long axis of ellipse should be placed at 90 degrees with core axis of circle.

Protractor

Used for determining angles of structures, so the information can be passed on to the man who bends or fabricates the specific part.

Dividers and Calipers

Used for checking equal dimensions or proportions on drawings or models.

Pentagon

1. Draw circle to given diameter of pentagon.
2. Draw radius CD perpendicular to center line XY.
3. Bisect CY, and with E as center and ED as radius, draw arc DF.
4. Use D as center and with radius DF draw arc FG.
5. Draw straight line DG. This is one side of pentagon.

Working Drawings or Blueprints

Final rigid drawings to scale on what to build and how to assemble.

Tracing Paper

A thin, tough, transclucent paper with a vellum (soft tooth) finish which lends itself to accurate pencil lines. The finished tracing (usually accomplished with pencils ranging in hardness from 2H to 4H) can then be placed over a light-sensitive paper like blueprint or brownline paper and exposed to light.

The exposed paper is then developed and the lines appear white against a dark blue or brown background. The original tracings thus remain clean in the designer's files, whereas the tougher paper copies can take thumbprints, spilled coffee, folding, mud, grease and footprints.

Tracing Cloth

A thin, transparent cloth permeated with a media that lends itself to top-quality inked lines. Used for the highest-grade precision drawings.

Specifications

Lists of materials, tolerances, and other parameters of assembly and operation.

Drawing Sizes

For standard filing practices and consistency between companies, drawings are usually made to certain overall sizes:

A size	8 1/2 x 11	or 9 x 12
B size	11 x 17	or 12 x 18
C size	17 x 22	or 18 x 24
D size	22 x 34	or 24 x 36
E size	34 x 44	or 36 x 48
Roll	Usually longer than 48 inches	

Computer oriented 3-D drafting machines seem to be getting too costly to set up & maintain. As a result, some manual 3-D drafting machines are being developed. One of these is the VERTEK, which can produce a perspective drawing from typical 2-D orthographics such as elevations & plan views. For more information write to Mr. Kermit Bowen, 9516 Bonnie Lynn Way, La Mesa, CA 92041.

Stylus's tracing these

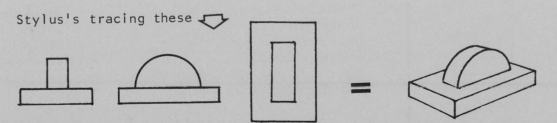

Preliminary Detail Drawings

When the designer and consultants are satisfied, the go-ahead is given, and parts, tolerances, joints, fasteners, and finishes are worked out tentatively in detail drawings. If it is not too compli- cated, the product can now be prototyped from these drawings. As a design becomes more complex, however, models may be made and several full scale prototypes. These will then be looked at by all concerned and discussed pro and con. Production men and sales- men are often in on the discussions, as well as the designer or mana- ger. Revisions will be made, and the item may be rebuilt several times until all the bugs are worked out.

The client may or may not be in on these early discussions and deci- sions, depending on the client-designer relationship. If the designer is part of the company manufacturing the product, obviously the owner or manager is in on all the steps. Whereas if the designer is a separate agency, the client may not be part of the discussions until the item is well thought out and presented as a unit ready to go. Several possibilities may be presented at the same time, but they are usually complete and within the area of solution as con- ceived by the designer.

In the early 1900's merchandising and designing brought in the idea that "The customer is always right." This was often mistakenly interpreted to mean "Give the client what he wants," and pleas- ing the client regardless of principle became, in too many cases, the not-to-be-disputed objective of designers just for the fast buck. As designers became more numerous, better educated, and more self-sufficient economically, they began to look upon themselves as professionals in the design area. This means that in many cases the client is NOT given choices in the design concepts or approaches. The designer examines the client's product, plant, and marketing situation, does the research and then proposes designs within the allowed budget or market environment. The client then accepts one or more of the answers presented to him, because that is why he hired a designer in the first place. Pleasing the client has a different connotation today. The professional client is pleased BEFORE he selects the designer or he wouldn't have approached him in the first place.

It is only the uneducated client today that hires a designer and then tries to tell him how to design. The obvious question is, of course: Why didn't the client design it himself in the first place?

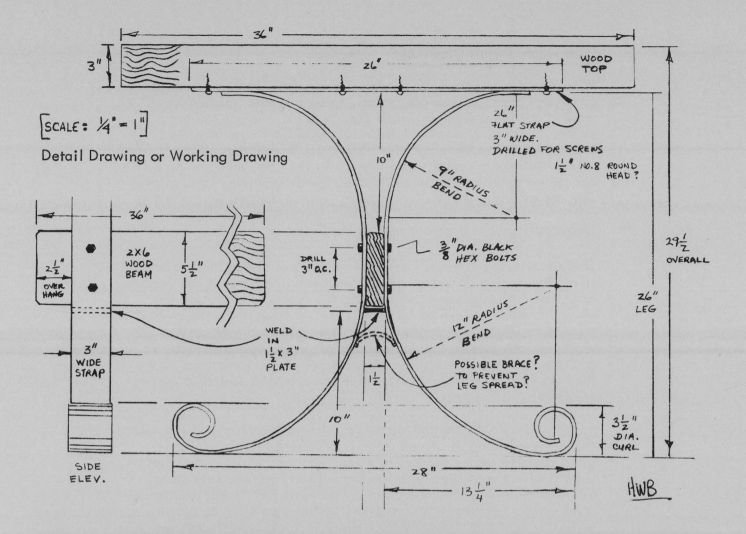

SCALE: ¼" = 1"

Detail Drawing or Working Drawing

36"

3"

26"

WOOD TOP

10"

9" RADIUS BEND

26" FLAT STRAP 3" WIDE. DRILLED FOR SCREWS 1½" NO.8 ROUND HEAD ?

3/8" DIA. BLACK HEX BOLTS

29½" OVERALL

26" LEG

36"

2×6 WOOD BEAM

5½"

2½" OVER HANG

DRILL 3" O.C.

WELD IN 1½" × 3" PLATE

3" WIDE STRAP

1½"

12" RADIUS BEND

POSSIBLE BRACE? TO PREVENT LEG SPREAD?

10"

3½" DIA. CURL

SIDE ELEV.

28"

13¼"

HWB

Prototype

Courtesy INCA Products, Inc.
Sante Fe Springs, CA.

153

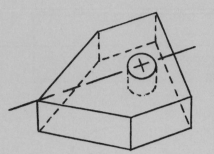

DRAFTING PRACTICES

When making detailed working drawings on tracing paper from which the blueprints will be made there are certain standard drawing procedures used the world over to describe 3-dimensional objects on a 2-dimensional surface:

EXPOSED or seen edges are shown by a SOLID line

HIDDEN or unseen edges are shown by a BROKEN line

CENTER lines are long & short THIN lines
(always starting & ending with a long dash).

CENTERS for holes are indicated by 2 crossed center lines with the short dashes crossing in the center of the hole.

ON CENTER is specified as OC. For example, four holes to be drilled 3 inches OC would be located as follows:

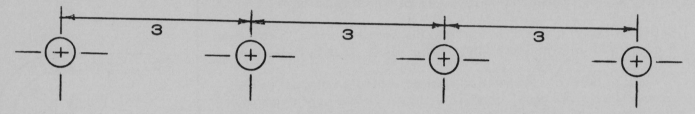

Holes are usually located from Center Lines of the plate itself, because the edge dimensions may not be critical enough to warrant precise tolerances. Thus below, A is better than B.

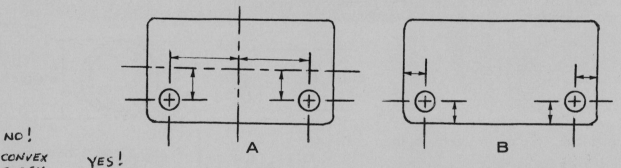

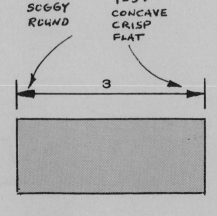

NO!
CONVEX
SOGGY
ROUND

YES!
CONCAVE
CRISP
FLAT

3

WITNESS lines are short, thin, extension lines which act as limits for the dimension lines.

DIMENSION lines are thin lines defining the distance between arrow-head tips.

ARROW HEADS should be 3 times as long as they are wide and have a flat or slightly concave back. In rare cases they can be used in perspective to help understand form.

BREAK lines are indicated by the following:
(Objects are not always shown full length.)

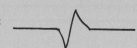

SECTION lines are bent lines that indicate where the object is to be cut or cross sectioned.

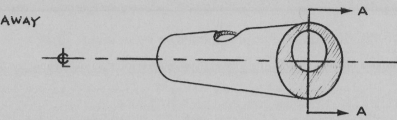

TOLERANCE (Tol.) is an amount which a measurement can vary over or under its dimension.

The figures, 22" ± .2, mean that the piece can be cut for assembly into the frame at any length between 22.2" and 21.8" but no more or less than that. The smaller the tolerance specified, the higher the cost usually goes. Tolerances of ±.001 require precision instruments, expert craftsmen and time, all of which raise costs.

RADIUS (R or r) is the distance from the arc of a circle to its center. A corner radius is indicated by a radius line. (See r.) ⟶

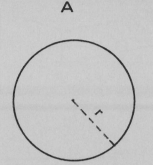

ANGLE of bend or construction must be indicated as well as the radius. Items like pipe, conduit or strap iron are bent over roller-like forms. The radius is usually indicated to the inside edge, so the bender does not have to subtract the pipe diameter or some other stupid dimension from his roller setting.

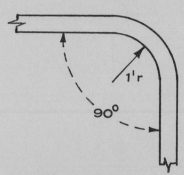

DIAMETER (Dia.) is the distance across the center of a circle. ID means Inside Diameter. OD means Outside Diameter.

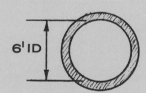
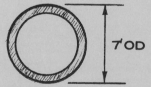

TAP means to cut a thread inside the hole or cylinder.

THREAD means to cut threads usually at the end of pipe, bolts, rods, etc.

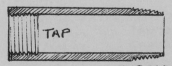

SPOT FACE means to grind or smooth the surface around a tapped hole so that the bolt head or washer can seat flat on the surface. Usually called out on coarse castings where the natural surface is pebbly or rough.

FILLET is a rounded portion that joins two surfaces at an angle to each other. Used for strength or esthetic reasons. Often the result of welding.

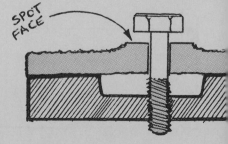

FAO-Finished All Over

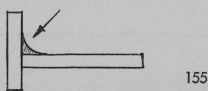

155

EXTENSION lines are the imaginary extensions of edges used to indicate relationships between views. Extension lines do NOT touch the drawings. Usually 1/8 to 1/4-inch space is sufficient to prevent confusing an extension line with a seen edge line.

PLAN VIEW   The top view. Looking down on or into the design.

ELEVATION   The side, front or back view. Flat, with no use of perspective.

Views should be in line. Thus if you were showing the plan view, front, and side elevations of a small boat, let's say, they would be related on your drawing as shown below. It is as if you stood in front of the boat, peeled the side on your right out to the right and peeled the top view up from the back, etc. Extension lines help relate views.

DEADRISE SKIFF

BALANCED RUDDER
1/8" GALVANIZED STEEL

CALL OUTS or LEADER LINES are done with an S-line or bent line. These will not then be confused with edge lines or other parts of the drawing.

LETTERING is usually in capital letters (Caps) between 2 fine, very light 4H pencil guide lines. After letters are drawn in leave the guide lines alone. Don't try to erase them. On most drawings, lettering 1/8-inch high is quite readable. Keep letters close together for better readability.

Condensing letterforms will help when you have to cram some information into a small space. Using expanded or extended letterforms plus wide spacing between letters will help when you want to stretch out a line to make a better composition or more interesting title block.

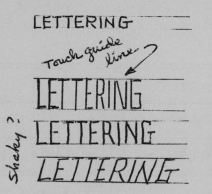

LETTERING

Touch guide line

LETTERING

LETTERING

LETTERING

shaky?

Roller coaster?

LETTERING

Don't go over it

LETTERING

LETTERING

LETTERING

LETTERING

LETTERING

space?   "oops"

LAYOUT

It is generally a good idea to draw a fine, dark border line around the edge of your tracing paper to define the edge. As you arrange or place your various views, detail drawings, and other specifications, try to allow a GREATER margin around the entire group than you have spaces between the items. This gives your layout better unity and keeps the items from looking as if they were going to fall off the edge. Also place as many of the items as possible in line with each other, just like you do with your plan views and elevations. Of course you don't need the extension lines as are required in the view relationships, but you can use implied extension lines by lining up edges, center lines, or sometimes corners to help the reader's eye move from section to section without feeling the items are a jumbled, chaotic mess. Artists often refer to this type of relationship as IMPLIED line.

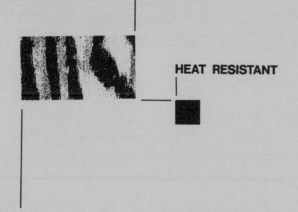

HEAT RESISTANT

TITLE BLOCK

This is the label for your drawing. It usually includes:

> Name of the company*
> Product title
> Designer's name
> Scale
> Drawing size and number
> Date of origin
> Date of any changes
> Draftsman's initials

*If you are retaining design rights, usually you do not put the client's name in the title block.

Title	TRESTLE TABLE		
Firm	INCA PRODUCTS		
Designer	Backus	Draftsman	HWB
Size	A	Scale	1/4
Date	Nov '74		
Change Orders	6/75		

9 PROJECTS

SECTION 9   PROJECTS

INTRODUCTION

With Section 9 we have arrived at a point where the beginning student can finally put his know-how to work on a "real" problem or assignment. He understands a bit about the role of the designer in this country. He is somewhat more aware of visual esthetics as well as the other design factors. Has a nodding acquaintance with raw materials and a few basic production and assembly principles. He may even remember to firm up his "Concept" before he clinches his "Approach."

So now, with the instructor's help, it is time to select a design problem and see if you can put it all together from first spark to finished sketch and possibly on to specifications and working drawings.

Then, if time and ability permit, a scale model or a working prototype may be produced. This last step is often dependent upon what the instructor and student decide is the best learning experience for that student at that time:

a.   Solving a variety of design problems on paper may be more necessary for a student who already has the ability to build models.

b.   Building a model or working prototype may be very necessary to those students who have never had to mock up a model or who have never had to work with or assemble metal components. Even one prototype often gives them a much better appreciation of how to draw or specify joints & fasteners. (How easy it is for the uninitiated to merely draw two pieces magically meeting at an edge without really knowing what they are indicating.)

Preliminaries

1.   As a first step the instructor may wish to designate a single, specific, rather simple project and have each student bring in a set of diagrams and idea sketches on this one unit. The sets are then pinned on the wall and critted by the instructor from the standpoint of communication. Legibility of line, written notes, logic of using different elevations, arrangement on the page, sequence of ideas, numbering steps, neatness, "dog ears," torn edges, sloppy presentation, and identification could all be part of the crit by the instructor. No grade is necessarily attached to this first effort, but the class learns quite rapidly that diagram or "sketch" does not mean unreadable slop.

2.   The sketches are held by the instructor, and either the same or another simple project is designated for the second round. When the second set of sketches and diagrams is turned in they are pinned above the student's previous work, and it is usually very apparent to all that the second sets look much more professional.

Product Selection

1. With the preliminary studies finished the instructor now asks each student to select two or three projects he might like to work on (possibilities listed on page 165 to page 198) and talk them over with the instructor. The student and instructor decide on ONE project, and set up a deadline for the first conceptual diagrams.

2. The student then turns in a set of diagrams and sketches which he feels contain the best concepts of his chosen product in the most communicable manner. (Later projects will probably be presented to the class as a whole for crit. But for this beginning round the instructor can usually speed things up and make all aware of standards more quickly by critting them individually.)

    a. Instructions to students regarding projects can be purposely very exacting with definite limitations and set specifications. Each student then is almost assured of success. The caution here is to be careful that the student is not so bullied into the final concept of the instructor that there is no student creativity allowed at all.

        However, this type of problem is probably more valid in the early part of the course, so that the student can firm up his sketching, line work, layout, sequence and general techniques of communication before he becomes too overwhelmed with a complex engineering or construction problem at the same time.

    b. At the other end of the teaching scale, instructions can be vague and "scrambled," so that the student has to limit, specify, or define his own exact concept through discussion and research. This "b-type" problem gives maximum amount of creativity to a strong student but might well lead to chaos, inadequate answers and finally indifference for a weaker student.

It is often noticeable that a certain type of student, A, will do very well with a-type problems. He draws well, specs well, and completes renderings with professional ability. Yet this same student may find that later in the course he has difficulty in coming up with original concepts. He dallies, he procrastinates, he asks instructors for concepts instead of technical advice and can just not create an original idea. Another student, B, less rigidly trained in drawing techniques, may do rather poorly on the a-type problems. However, later he may discover he has original ideas or creative modifications, regarding the b-type assignments, coming out of his ears. And if he works hard and picks up the drawing and rendering techniques as he goes, he often finishes the course in a flurry of success.

But A, our first student, finds it much more difficult to pull himself out of the morass of cliche', tradition-bound thinking. Hand him a concept to copy and present, and he does great. Give him a b-type problem, ask him to originate a new idea, and he chokes.

So remember, young designers, it is not enough to learn to draw and render or photograph and arrange. These are just tools. You must keep your education going on all fronts. Art history, sociology, travel, architecture, physics are all necessary for your background. A limited, narrow education of technique application only has never produced a designer. If you A-type students, in particular, want to help yourselves grow as designers, try solving as many design problems as you can each day. At the end of this section there are recommended exercises in design. Some are simply a matter of discussion and applied thinking. Others are more complex and require some research before solutions are possible. Whether you come up with an ideal solution to each problem is not the criteria of success. What is important is the constant practice of developing your mind and self-assurance by tackling one problem after another, day after day.

The problem-solving exercises in the above paragraph are just as necessary for you great renderers who cannot originate, as drawing and rendering exercises are necessary for the creative student who cannot communicate his ideas properly. Right?

Design involves the following thought processes:

1.   Formulating the problem. Asking the question. Defining your concept. How used? Fits what market? Who uses? Who buys?

2.   Listing the limitations. Cost? Materials? Size? Weight? Deadline? Power? Etc.

3.   Organizing present data. Existing components. Competitive products. Past assembly procedures on old design, etc.

4.   Defining possible solutions. The sparks.

5.   Considering production possibles. Synthesizing various approaches.

Formulating the concept, defining the problem exactly (step 1 above), is probably one of the most important steps in the design process. Albert Einstein said, "The formulation of a problem is far more essential than its solution, which may be merely a matter of mathematical or experimental skill."

There was a weird family named Stein.
There were Gertrude, Ep and then Ein.
Gert's writing was hazy,
Ep's statues were crazy,
And nobody understands Ein.

DESIGN STEPS

In any one of the suggested problems in this section the student could take it as far as he and the instructor decided. Following are 6 steps that seem to provide reasonable stopping places. Step 6, manufacturing the product, is pretty far fetched for student products, but it has been done. This list of steps is another means of letting the student see exactly what is involved in communicating a creative idea to someone else.

1.   Concept diagrams & sketches

Getting possible solutions down on paper. Not attempting any particular approach or specification yet. Brainstorming. Try even the ridiculous. Worry about feasibility later.

2.   Approach drawings and illustrations

Choose one solution, firm up the drawings, specify materials and dimensions so that student and class can understand and discuss the exact design from both market and engineering viewpoints.

3.   Complete presentations

Assuming the idea will now be presented to a client, make 3/4-view Canson paper or other type renderings with elevations, plan, or cross-section views to scale where necessary. (See Section 8.)

Material lists and costs plus copy regarding marketing or manufacturing information may be necessary now also. In other words, a total presentation, so a client would have all the information he needed to make a decision.

4.   Building a model to scale

This model may or may not be a part of step 3. Materials such as balsa wood, cardboard, contact paper, conduit, welding rod, plastic pipe, tongue depressors, popsicle sticks, soda straws, adhesive tape, body putty, etc. can all be used to simulate the structure. Spray fix can provide a surface over which one can paint enamels, latex paint, or lacquer. Shellac makes a good undercoating in some cases also. (See Section 8.)

5.   Build an operating prototype, full size

If the product is a rather complex appliance or other machine, this should be attempted only by the student who has had some basic experiences in metalworking, plastics, and electricity. A problem of this nature often requires several months of individual work. And it is always a good idea for the instructor and student to discuss the job together, so that both are certain there will be meaningful learning experiences for the student in that period.

6.   Produce the product

There are two ways the designer can go here:
a.   Sell the idea to a company. Let them manufacture it and pay you a royalty. (See notes on royalty contracts page 236.)
b.   Manufacture it yourself. (See story regarding two young men and their beginning business on page 195.)

TIME SCHEDULES

The following outlines are items to consider when planning a design schedule. It will help you, the beginning designer, to glance over these a few times while you work to keep first things first. The two schedules overlap somewhat. For example, sharp conceptual sketches or design drawings may sometimes be used as part of the presentation program to a client, etc., but the two areas of inplant design time and the (often separate) time scheduled for presentation to an outside client are usually two different ball games.

Design Time Schedule

Budget your design time so you won't miss deadline.
Work backward in time from the deadline date. (You will probably be surprised to find that the first sketches should have been done yesterday!) Then allocate the following general steps a specific time span.

1. Search for concept. Brainstorming sessions without worry about feasibility.

2. Conceptual sketches limiting ideas. Feasibility IS considered now. Competitor's products considered also.

3. Modification drawings. The approach. Existing parts or components are considered. Cost, simplicity, nuts & bolts, assembly, etc.

4. Development, partial mockups, variations, testing.

Presentation Time Schedule

1. Renderings and overall specs. Some explanation and copy.
2. Model making
3. Fotografs. Slides. Movies.
4. Talk or tape.
5. Brochures or booklets (detailed information and specifications.)

163

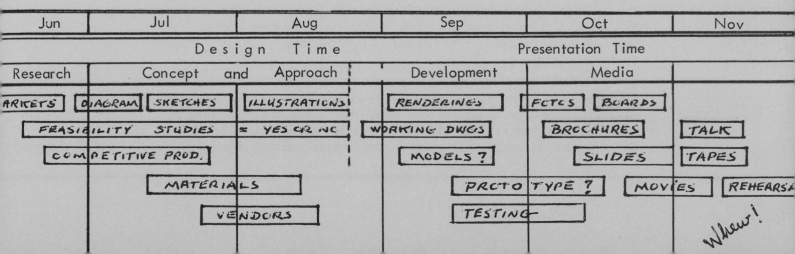

Jun	Jul	Aug	Sep	Oct	Nov
Design Time			Presentation Time		
Research	Concept and Approach		Development	Media	
ARKETS	DIAGRAM SKETCHES	ILLUSTRATIONS	RENDERINGS	FCTCS BOARDS	
FEASIBILITY STUDIES		= YES OR NO	WORKING DWGS	BROCHURES	TALK
COMPETITIVE PROD.			MODELS ?	SLIDES	TAPES
	MATERIALS			PROTOTYPE ?	MOVIES	REHEARSA
	VENDORS		TESTING		

Whew!

EXCURSIONS

Although this is a book with the emphasis on originating, graphics, and presentation — the drawings or models should not necessarily be considered the end result of a student designer's effort. Actually, the graphics are only one factor (although important) in the communication chain which would end in manufacture. As designers realize, producing the actual product entails a whole new set of problems for the young designer. An older, experienced designer will have foreseen most of the production possibilities and worked within them. As a student designer, try to visit one manufacturing or assembly plant per month in your area. A furniture manufacturer, a toy company, plastic-forming plant, a welding shop, steel mill, stamping mill, etc. will all add to your production knowhow. Larger companies like steel mills, auto assembly plants, or paper mills often have regular guided tours. Smaller companies should be contacted ahead of time. Sometimes a product design office will not encourage visitors as they are afraid of "pirating" by competitors, but mentioning you are a student will help. Process plants, like cladding, engraving, welding, carpentry, printing, sheet metal, shoe repair or formica are usually happy to see you if you let them know a week ahead of time.

A product does not necessarily have to be mass produced to help a student understand production. Building a single "working" model will often give the student a real experience in metal bending, welding, drilling, fitting a piece of glass in a "window," fitting a knob, etc.

There is the classic problem of making a clay model of a tool handle, say for a screwdriver, spray gun, drill or perhaps some other small device. The student follows through by casting a two-part plaster mold around the clay model. The mold is then used to cast a resin prototype. The young designer learns a number of things about plaster mixing, vents, releasing agents, shrinkage, flash marks, precision, plastic catalysts and finishing processes.

However, there is one disadvantage of this type of technical problem, particularly in the junior college or lower-division design classes. It is the length of time involved. The student might have been better off if he had spent the same 6 to 8 weeks working on quick visualizations (both 2 and 3-dimensional) of a variety of concepts. In advanced design schools there is usually an open lab where the design student can work at anytime during the day and evening. He thus gets a number of chances to become fairly proficient in technical processes and producing a working prototype at a later date.

There is no "correct" answer. But the student in the beginning product design classes and the instructor might well discuss the objectives of a very lengthy technical process problem. Whereas the product design classes in the upper-division areas or advanced design schools are usually filled with students majoring in product design, there are many students in the beginning lower-division classes that may not have settled on product design as yet. They might benefit more from shorter-length problems involving a greater number of design decisions.

SUGGESTED PROBLEMS

The following problems may often spark other ideas from students. Unless the problem is a definite assignment by the instructor, the student may well think up a modification or direction change in the basic objective. Anything that encourages a student to think on his own, even though his beginning concept may sound ambiguous and hazy, should certainly be part of these design studies.

Students can learn by not quite solving a tough problem in the same manner they learn by solving a simpler problem. The continued effort of being curious and seeking solutions helps foster creativity. The introduction of uncertainty into a problem tends to give a feeling of uneasiness (particularly to the instructor, who is well aware of the probabilities of failure) to those involved. However, if the student is encouraged to feel he is thinking on his own and is being an original person, he will undoubtedly benefit by the experience, even though his final answer is pretty much of a flop.

Also there are many problems in this world that adults know the immediate answer to. But sometimes it helps the student to define the problem anyhow and seek out the answer by himself BEFORE we give him the "Mission Impossible" flag. The definition, the effort, the thinking are what count. Once he gets his character set on the METHOD of seeking solutions, he has it made.

MATERIAL EXPLORATION

Design an item that is functional, unique, simple, and saleable, with a minimum modification of material.

Perforated aluminum
Expanded metal
Contact paper
Paper laminates
White formica
Acrylic clear plastic
Foam core
Plastic cylinders
Tin cans
Bamboo
Welding rod

Celcon plastic
Corrugated cardboard
Canvas
Milk cartons
Piano wire
Redwood slat snow fences
Lumber 1 x 2's, etc.
Doweling
Plastic pipe
Chimney flue tiles

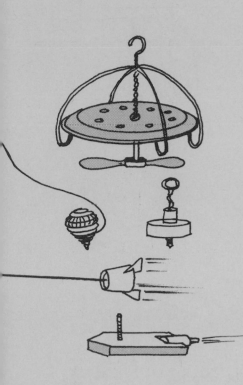

TOYS

Using articles found around the house, design several toys that move. Build one and demonstrate it to class. Some possibilities are:

Hovercraft

Aluminum foil pie plate, plastic propeller, coat hanger, and heavy rubber band might do the job.

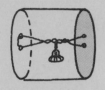

Tops

Handspun, string-whipped, spring-loaded, or twisted helical strip are a few possibles.

Prop cup

Plastic cup, rubber bands, sticks and cardboard for rudder or stabilizers?

Soda boat

In grandpop's day kids used to use baking soda and vinegar to propel tiny wood boats. Can you?

To & Fro

Rubber band stretched between the ends of a cylinder and weighted in the middle with a lead sinker and paper clip causes the rolling toy to roll back after it is rolled away.

Balloon on wheels

A long balloon mounted on wheels with a control valve in its opening will sputter along the playroom floor for quite a while. No valve equals rocket.

Skidoo stick

A 3/4-inch square stick about 8 inches long with notches cut along one corner. A well-balanced propeller about 3 inches long nailed into the end. Hold the stick in one hand and scrape a pencil or small stick along the notches in one direction. The propeller will gradually start to turn. Then shout "Skidoo," scrape the stick along the notches in the other direction, and the prop will reverse its rotation. The prop must be well balanced.

Use left-over model parts to build your OWN design.

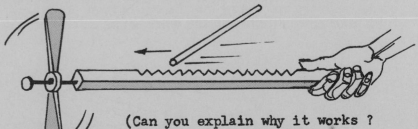

(Can you explain why it works ? Neither can anyone else.)

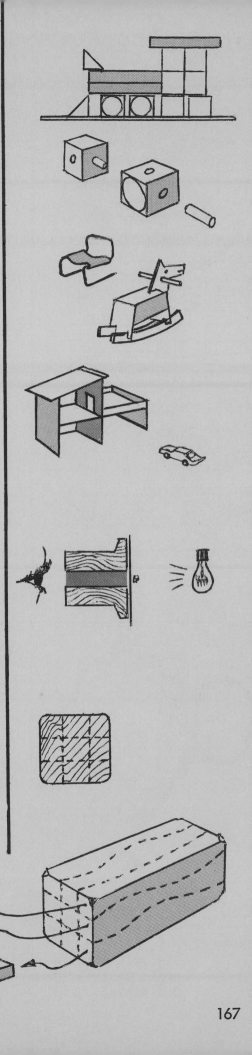

Module blocks

Sanded wooden blocks cut to precise module dimensions will provide a variety of transportation or construction machinery for the creative child.

Blocks & dowels

These same module blocks can be drilled on all sides so that slip-fit dowels can be used to hold the assembled toy together.

Paper-clip furniture

Miniature doll furniture can be made with paper clips and scraps of cardboard. Quick-drying model airplane cement or bits of tape help assembly.

Cardboard rocking horse

Cardboard dollhouse

Spool microscope

Microscopes are dependent on lenses. But you can make a microscope by shortening the focal length of your optical system also. Paint the inside of a wooden thread spool black. Glue or tack a piece of clear acetate over one end. Place objects like grains of salt or a butterfly wing on the acetate and look thru from the other end at a strong light. If the magnification is not very large, you may have to cut the spool down in length.

3-D jigsaw puzzle

Find a piece of soft wood, like pine, about 3 inches square and 8 or 9 inches long. Sand it down. Then cut it into nine pieces lengthwise on a band saw by curving the cuts as you feed the wood through the saw. Stain and linseed oil the pieces and you have a three-dimensional jigsaw puzzle that is a bear to put together. The rounded corners on the four corner sections are a start but it's still a tough solution. Lay it on your coffee table and see what happens.

TOOLS

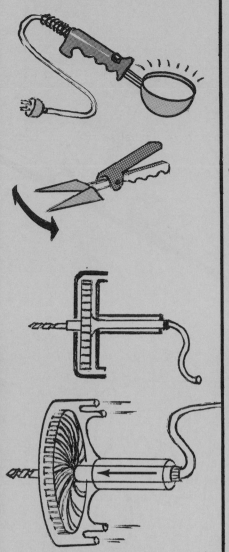

Ice cream scoop with quick-heating electric element in edge of cup.

Garden clippers: Develop a simple mechanical system to translate up and down motion of hand to side cutting motion of shears.

Multipurpose tool for housewife: Include screwdriver, pliers, shears, wrench(?)in one unit.

Multipurpose camping tool: Include shovel, hoe, axe, pick (?) on one shaft.

Underwater hydraulic drill: Use pumped water to operate turbine or other driving device. Exhaust water must be used to propel drill forward, because floating diver oftentimes has no way to push against work forcibly enough to give drill a bite on material.

APPLIANCE

Module Toaster

The following diagrams and sketches have been given you by a client. He wants you to draw up the illustrations so that the idea of interconnecting modules is self-explanatory. He would also like a drawing of one of the inside heating-grid panels showing the insulation supports and circuit wires from the timing mechanism through the heating grid panel back to timer or bypass. There will be no "pop-up" mechanism, no light-to-dark setting or bottom cleaning trap. The client wants a very simple, foolproof toaster without all that interior maze of wires and mechanisms that accumulate crumbs, grease,and dirt and defy repair in present toasters.

TIMER

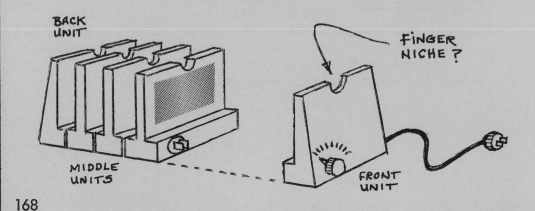

BACK UNIT

FINGER NICHE ?

MIDDLE UNITS

FRONT UNIT

SPORTS

Simple Ski Binding

New type ski binding. A simple, inexpensive binding for those skiers who enjoy skiing over rolling country with gentle slopes. There are many skiers in the Midwest and northern countries who have skied all their life and never saw a ski jump or a slalom course. It is rather ridiculous for these kids and adults to buy a $110 pair of boots and a $70 set of ski bindings when something simpler would be MORE efficient. . . for their purposes and easier on the ecology.

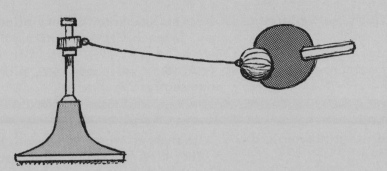

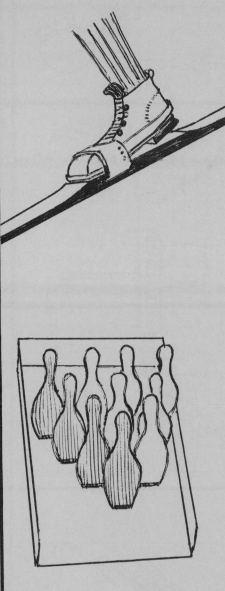

Indoor Tether Ball

Smaller ball, shorter rope, on a wide, nonskid base might be fun in the playroom. Ball could be hit with paddles perhaps, instead of hand?

Indoor Bowling

Flip-down, flat bowling pins made of masonite, plywood, or equivalent, to be hinge-mounted on a flat board which ramps down to thin forward edge. Weight and size of pins would depend on weight and size of the bowling ball used. Ball should probably be selected first from balls already on market. A heavy ball is necessary, as the lighter, air-filled types would bounce off the pins.

Can you design pins so that if ball hits exact center of No. 1 pin all pins fall? But if it hits to one side only some fall? Do they fall back into sockets flush with board so ball doesn't bounce over hinges? Etc.

Expandable Camper

Design a small camper trailer which can be expanded by pulling out nested modules. A type of waterproof seal at joints to be worked out. Lightweight honeycomb walls might be one answer. Could it be used in the backyard as storage space in off season?

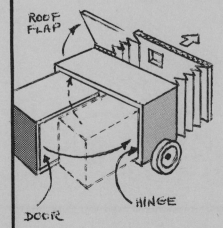

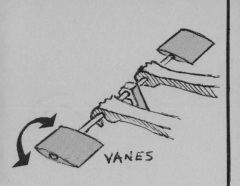

VANES

A "BALANCED" RUDDER MEANS THAT PART OF THE SURFACE IS AHEAD OF THE AXIS IT TURNS ON.

THUS THE RUDDER OR VANE CAN BE HELD IN POSITION WITH LESS FORCE, BECAUSE THE WATER BEING DEFLECTED BY THE FRONT SURFACE HELPS TO BALANCE THE WATER PUSHING ON THE BACK PART.

Underwater Bicycle

Sketch out or diagram a few ideas for a self-propelled underwater vehicle using the bicycle pedal ideas as the driving force. A propeller could be driven by chain or gear train. Vanes in front part of device could be controlled by the hands to direct the vehicle up and down. A rudder at the rear could provide left-to-right movements. This would probably have to be activated by the hands also, because the feet are driving the pedals for power. Do you think a flywheel in the drive train might help modify fluctuations in power or provide a bit of "freewheeling" power while the rider rested his thighs occasionally?

SCUBA (Self-Contained Underwater Breathing Apparatus) might be attached to the vehicle frame. Is the vehicle small enough to fit into a car trunk? Does it have to be foldable? Would there be any particular problem with the bearings or gears after being immersed in water? Should they be sealed or left open so water can be blown out before storing, or what? What type of "saddle" would be necessary for the driver to lie on? Is a strap or hook-on arrangement necessary, or would the hand grips be sufficient? Turning the vanes perpendicular could provide braking power.

AXIS

BALANCED RUDDER

MAST

BOOM

PIPE

Land Sailer

Bicycle wheels, mast and boom, pipe, sail and seat can be combined for a lot of fun whether you sail it on the hard sands of the California deserts or empty parking lots on Sunday. On the larger models, consider telescoping the frame for ease of towing behind car.

DIHEDRAL ∠

Kites and Gliders

Stability may be achieved by dihedral angle, weights under wings, flexible tails that act as dampeners on yawing or pitching, as well as tabs, rudders and elevators.

TILT MEANS LESS LIFT ON LEFT WING.

Bobsled

PLANE RETURNS TO HORIZONTAL.   170

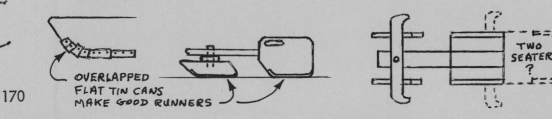

OVERLAPPED FLAT TIN CANS MAKE GOOD RUNNERS

TWO SEATER ?

Kids Volleyball

Little kids don't have much fun playing volleyball, because the ball is so heavy it hurts their wrists and hands. Small children complain about it and cringe when a high ball arcs down upon them. Yes, there are light, plastic beach balls on the market, but they are too bouncy or too soft to give the volleyball feel.

See if you can design a volleyball light enough so it can be used by the little kids as well as muscular people. Anyone can play a sport if he is strong enough to control the stick, bat, racket, ball or bow. But doggone it, when you are born weaker and smaller than Amazons and top athletes you should not be expected to handle the exact same weight and size of equipment that was designed for the pro. In fact, this is a rather stupid tradition that has taken the fun out of many ball games for many kids. In archery one is given a weaker bow. In golf one is given shorter clubs. But in volleyball or basketball, the little hand is supposed to compete equally with the big hand. Well, 95 percent of the kids are NOT going to ever play anything near pro ball. Why not, for Pete's sake, design equipment to suit THEIR recreational needs, and have it evaluated by the kids themselves, or at least a sensitive recreation coach who isn't a pro himself.

If you are designing equipment for professional use then use professionals to evaluate your results. But if you are designing for the average person, let someone evaluate your results who deals with average people.

If the ball is too light, it will act like a ballon. That is, it will not fly over the net when struck. Suppose you put three balloons inside of each other and blew that up? Or, could you start with a rubber bladder and sew up an outside casing using a variety of materials until the ball was just right for the kids? Maybe build a case for one of the small beach balls?

A regulation volleyball net is 8 feet high, which is about as high as a 5-ft.,10-in. person can stretch reach with his finger tips. Plan a net with a height related to the finger-tip reach height of the kids you're going to test your ball on.

Then get, say, 4 kids on a side. Let them test out the various balls and see which one they think makes the most fun game.

How would you manufacture it? Want to visit a ball manufacturer like Voit or Spaulding and find out how they do it?

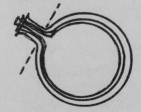

Tennis Ball Redesign

Tennis balls are notorious as throwaways. They deteriorate because of two factors: the pressure inside leaks out and they become dead, or the fuzz wears off the surface, and they float or wobble as they fly thru the air. The pressure problem has been solved by several firms, such as Tretorn of Sweden, by making a pressureless ball which has a thicker more resilient wall, but the wearing off of the fuzz on the cover still prevents them from being used after a few sets.

Could you peel the covers off a few pressureless balls and re-cover them with a piece of tough material which had holes or grooves drilled thru it? Save the cover you tear off and weigh it so you can use a new material about the same weight to keep the balls within specs. Adhesives such as GE's Silicone Seal or gasket cements might be a start. Make different size holes and space them differently on each ball you cover. Then get a hold of a pretty good tennis player, an amiable pro preferably, and let him compare your balls with regular balls as regards topspin, slice, chop, etc. The point being that if you could find a reasonable answer, you would have a tennis ball that could be used for maybe twenty sets before the cover became worn so thin that the holes will not act aerodynamically.

The idea, of course, is that there is no reason small holes in the cover of the ball wouldn't take the place of the soft fuzz. That is, the holes would act aerodynamically in the same way the fuzz acts to keep the ball from floating or wobbling and would not wear away anywhere near as fast. Jeepers, think of the thousands of tons of rubber and synthetic nylon fabric that could be saved for other purposes. Now you will naturally have critics saying, "It just won't work, because the fuzz is necessary to grab the strings of the racket. And besides, the ball will skid when it hits a smooth court surface. And besides, the manufacturers WANT to have people throw away the balls, because this gives them more sales, and besides.............ad nauseam." Illigitimi non carborundum.

You should remember again, that the great majority of tennis players are NOT pros or even good club players. I have played tennis all my life—in fact, I have written a book on it ("Checklist for Better Tennis," paperback, $1.25, Doubleday.)—and many are the aspiring players I have watched batting back & forth a single, dirty, smooth, dead tennis ball because they couldn't afford a three-dollar can of new balls every other day! Wouldn't it be great if you could convince the Tretorn company in Sweden to develop a ball like this, EVEN IF IT WEREN'T ACCEPTED by the International Lawn Tennis Association immediately? Think of the millions of people that could benefit by having a tennis ball that wouldn't wear out after twenty minutes of hard usage.

Tretorn core wall

Regular core wall

Covers could be prepunched and then glued over core without any major change in assembly machinery.

One last suggestion: Getting the covers off a tennis ball is one hell of a tough job. Tearing them off with a needlepoint pliers or trying to sand them off on a sander is almost impossible. So how about contacting the Tretorn company's rep in your own country; explain the nature of your student project and see whether they would send you a dozen pressureless cores without the covers so you could experiment with them? If you were at all successful you might convince them to manufacture some. And you have made a contribution to ecology, eh?

Workup Table Tennis

Design a round Ping-Pong table about 6,7 or 8? feet in diameter. Cross two nets thru the center at 90°. (See diagram.) Nets will be 6 to 10 inches in height, depending on what you decide after playing a few experimental games. Regular Ping-Pong paddles and balls might be a start, although it may lend itself to a heavier or slower ball patted with the hand.

Four can play with as many as four or five waiting in line for court 4 as soon as a point is lost. The player at court 1 is the server who starts the game by serving, usually Ping-Pong fashion, into any one of courts 2, 3, or 4. (A let serve is played over. A let on any other ball is played as usual.) After the ball bounces in the court, the receiver may hit it back on the fly into any one of the other 3 courts.

If the server faults he must go to the end of the waiting line. Player in court 2 moves to court 1 and becomes the server. Player 3 moves to court 2, player 4 to 3, and the person first in line goes to court 4. If any player besides the server makes an error like netting a return, or knocking the ball out, he goes to the end of the waiting line; the server gets a point; and all players move up to fill the court left open by the player who erred. The server continues to serve and is the only one who can make points (like volleyball). And he gets a point every time someone is made out, even if he didn't hit the deciding ball. When the server is made out he keeps his points and can accumulate more if and when he works his way back to the server's position again.

It is also an out if the ball touches any part of the player's body. Sometimes a player will attempt to "slam" the ball into one of the other players. The player so hit is out, but of course, if he dodges and the ball misses, the slammer is out.

The last player in line is the umpire. But he does not give a decision unless appealed to by the players involved.

A tennis ball must be more than two and a half inches but less than two and five - eighths inches in diameter. It must weigh at least two ounces, but cannot weigh more than two & one-sixteenth ounces. When dropped 100 inches to a concrete base, it must bounce at least 53 inches but less than 58 in.

Ump

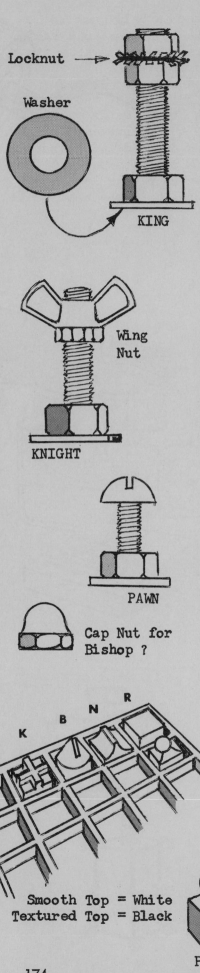

Locknut →

Washer

KING

Wing Nut

KNIGHT

PAWN

Cap Nut for Bishop ?

CHESS

Nuts & Bolts Chessmen

Design a set of chessmen by using bolts, washers, nuts, etc.

These can be attached to their bases by welding, soldering, or possibly using a strong industrial adhesive or catalytic resin glue. (This last usually comes in two containers which you pour together when you want to use it.) If you decide to have the pieces chromed or brass plated, you had better check with the foreman of the plating plant to make certain the solder or industrial glues you used won't come apart during the electrolytic process or whatever happens in a plant like that. Maybe you should call up the plant and see if you could go over and get a tour someday. Tell the manager you're an eager-beaver, budding designer and would like to understand the plating process better. To make your tour more meaningful go to your library; get out a recent book on plating techniques and read up on it as fast as you can. Skim it for general information. Glance over the table of contents. Read topic sentences. (If you will remember your old English teacher, Mrs. Sedgewick, she told you that quite often topic sentences occur at the beginning of paragraphs. They are the sentences that sum up the content of the paragraph.)   Read as fast as you can until you get a general knowledge of the processes and some of the limitations. Then when you go to the plant you will be able to ask some intelligent questions and suck more information out of the shop foreman, or whoever takes you around. If you feel bashful, take a friend with you.

If you don't play chess, be certain to talk to an old chess hound about sizes and appearance of the pieces. There are many exotic, spectacular chess sets in the world, but ninety percent of all real chess games are played with rather standard pieces which follow a certain precedent set by the Staunton Chessmen from England. The height of the pieces is in some relation to their value, they can be handled easily, the bases are weighted for stability, they are readily identified, and the difference in color or value between "White" and "Black" is in relation to the color or value of the chessboard squares.

Chess for the Blind

Design a chess set for the blind. The board would probably have to have sockets of some nature to receive the chess pieces so they would not be knocked over when the players run their hands over the tops of the pieces to determine the positions. Also, the indentations, or "sockets," must be obvious enough so that their hands can determine the diagonals, ranks, or files along which the action takes place. Tops of the pieces must be designed with significant forms so the players can determine which piece is which. Height may or may not be of importance. It might be wise to consult a blind person before settling on your design concept, regarding size, shapes, loose pieces on the side of the board, etc. (Well, then learn to play chess now.)

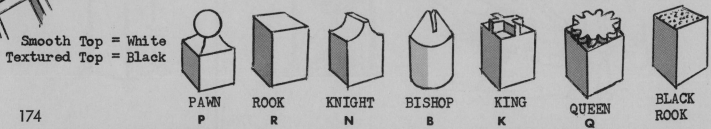

Smooth Top = White
Textured Top = Black

| PAWN | ROOK | KNIGHT | BISHOP | KING | QUEEN | BLACK ROOK |
| P | R | N | B | K | Q | |

THE HANDICAPPED

Utensils for the Handicapped

Design a knife, fork and spoon for those persons who have slotted steel hooks for hands. Consider ease of picking up from the table as well as use. Again, a person with that particular handicap should certainly be consulted.

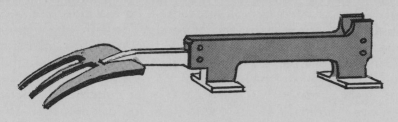

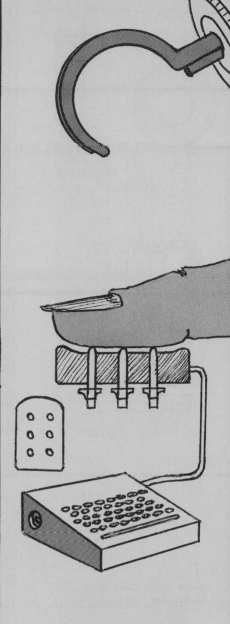

Braille Receiver

The American Foundation for the Blind in New York City considered a machine at one time whereby a plate the size of a finger tip had 6 holes punched in it through which 6 rods would protrude in certain patterns to simulate the six embossed bumps in the Braille system for identifying letters. The rods were activated by 6 solenoids (electro-magnets) under the plate. The solenoids were in turn operated by a typewriter keyboard set on a small box with switches inside (powered by three flashlight batteries) which energized the solenoids when the keys were struck. Thus, a blind, deaf person could place his finger tip on the plate and read the message typed to him by his companion. Would you like to design such a system? Or can you think of any other application of such a device?

Recycled Gift Item

Design an item for a gift shop from readily obtainable or even throw-away materials that could be easily assembled by handicapped persons. Don't try to compete with mass-produced articles already on the market. Rather design one expensive item that is to be handcrafted (custom made) by the handicapped individual and sold at a good markup through gift shops.

WIRE INSECTS

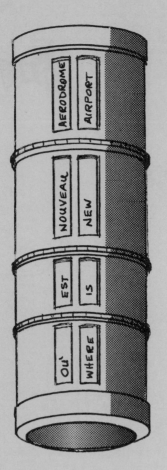

Translator

Can you visualize a rod about six or seven inches long, maybe two inches in diameter with four or five movable drums or collars on it? On each drum is printed about 12 English words with a foreign language equivalent next to it, so that when each drum is turned to expose the English word through a slot, the foreign equivalent is exposed in the slot above it. This becomes a translator then that could be stuck in the pocket and used in emergencies in foreign countries. Drums might be interchangeable so different language drums could be slipped on the rod for different countries. One drum might include personal pronouns, two might have a variety of verbs, with adjectives or adverbs on the fourth. The last drum might contain a variety of nouns important for tourist "survival."

Pronouns?	Verbs?	Adjectives?	Nouns?	
I	want	hot	food	taxi
me	did	cold	sleep	toilet
you	didn't	hard	doctor	theatre
he	do	soft	police	church
she	don't	clean	hospital	clothing
it	can	dirty	ambulance	water
they	cannot	new	airport	bank
this	will	old	train	money
who	was	fast	bus	travel
———	is	slow	today	man
what	am	wet	tomorrow	woman
why	not	dry	yesterday	boy
where	etc.	here	ticket	girl
when		there	passport	etc.
how			Am. Express	
etc.		etc.	auto	

The above lists are examples only. You, the designer, may find a different set of words or fewer drums much more effective. One way to test is to imagine different emergency situations and see if the Translator will help you solve it.

OR

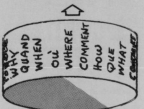

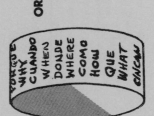

Where is new airport?
I am sick. (Ooops – no word for sick?)
When does bank open?
I lost my passport.
We sent you letter.
Is this local bus?
Is this express bus?
This is first-class ticket.
Where is American embassy?

Adjectives are probably the least important.

LIGHTING

Floor Floodlight

Use tubing or square conduit or equivalent material and mock up
an adjustable balanced floor lamp which might be used by artists
or photographers to light a variety of subjects from different angles.
If you design for high wattages, then consider socket cooling through
slits or holes so that shade or reflector doesn't "smoke." Balance
weight should probably be adjustable so it can compensate for variation in lamp shade weight. Swivel should have adjustable clamp so
it can be tightened and loosened by hand with a minimum of struggle.
Length of levers is important. Base must be heavy enough to prevent
lamp from tipping over at the slightest bump. If vertical standard is
screwed into base there should be some locking device to prevent it
from unscrewing as lamp is swung into different angles. Where does
cord come out? Suppose light is swung through a 360-degree circle.
Will the electric cord then be twisted off inside the conduit? Can
you use stops? How do you propose to allow the photographer maxi-
mum leeway in adjustment with minimum inexpensive electrical
contacts?   Maybe you better go look at some lamps first.

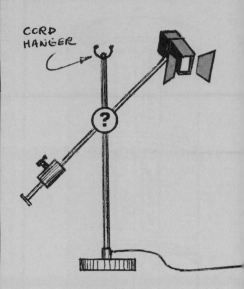

Hanging Lamp

Design a hanging lamp for general illumination. Cut out five to
seven identically shaped pieces of stiff cardboard. Tab, slit, glue
and interlock them in an asymmetrical way to encompass a light
bulb. Make certain that you have a variety of negative spaces to
act as a foil (opposition) for the symmetrical positive shapes. Consider the silhouette form from all sides also. Once the form is fairly
well jelled, consider different materials you might use to build it.
Clear plastic, stained-glass, perforated masonite or aluminum might
all be possibilities. How is light bulb mounted? Where does light
cord come out. Where is center of gravity when it is hung? Does it
hang level? Does it throw light upward as well as downward?
Sideways? How is cord attached to ceiling? Is it strong enough to
support lamp or will you need chain? Etc.

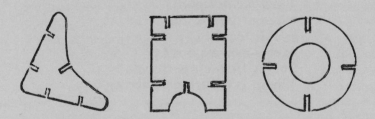

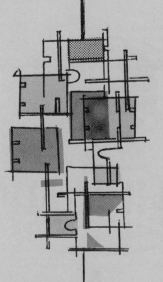

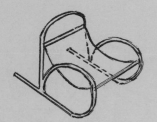

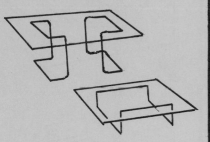

FURNITURE

Chair Frames

Using a rather stiff heavy wire, simulate several tubular chair frames. How will the model support seat , back , arm rests? Will you envision wood, canvas, or lacing for backs and seats?

Three-in-One Module

Bend paper clips into several configurations to simulate tubular table legs so that one configuration would allow for three different table heights when placed in different positions under the table top. Two of the same configuration may be used together, if necessary.

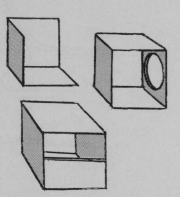

Slot and Tab

Design a set of furniture for a child's room using flat surfaces, cut in various ways to allow for slot-and-tab assembly.

Cube Storage Wall

Build an interesting storage wall by using one, two, or three modules of a box form. For example, a cube could be open one side, open two sides, half side, three sides, circle opening, or whatever. However, too many different modules run the cost up. Build the storage wall, then, by stacking scaled-down cardboard models, or it could be shown in a perspective drawing.

STEP CHAIR

The Senior Designer has just given you the diagrams & sketches below. Design a chair which when folded forward, as indicated, will become a small stepladder. Make up a parts list: kind of wood, pieces needed, hinges, braces, screws, doweling, etc. Locate a woodworking shop in your area and call the manager to see if you can get a half-hour interview with him. Tell him this is a student project and see if he would look over your parts list and prices and help you estimate what the manufacturer's cost would be plus markup to the retailer. Labor, materials, overhead and profit margin would be included in the price to the retailer.

Bring your figures to class. Put your own drawings on the wall and let each student in the class make a quick estimate of the four cost factors on a separate piece of paper. Labor, so much; materials cost; overhead costs per chair; and markup to retailer. Then post your figures from your interview and see how they compare with other students' estimates. This makes for some surprising results and interesting discussions regarding businessmen's problems of staying solvent.

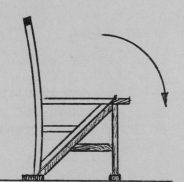

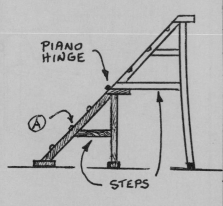

PIANO HINGE

STEPS

Ⓐ

HOLES AND PEGS TO PREVENT SIDEWAYS RACK WHEN CHAIR IS IN NORMAL POSITION?

CHAIRS

Although a chair seems to be simple enough object, there are literally thousands of different designs. From church pews, which are designed to keep one upright, to canvas sling safari chairs that invite the body to sprawl horizontally, the designer can choose from an almost infinite variety of sculptural forms and materials.

The chair takes abuse, so it must be designed with regard to rack. That is, do the joints, fasteners and its structure keep it from collapsing sideways, or fore and aft? Can it take a 200-pound body weight dropping down on it from a height of one foot several times a day or more?

Does it look interesting? As a designer are you aware of the spaces (negative areas) under the seat, under the arms, etc., as well as the thicknesses and color of materials used? What about overall silhouette form, especially the front and side elevations? The three-quarter view? Are you certain an 18-inch by 18-inch seat is sufficient to hold a modern day butt in a lounge chair? If it is deeper will the front edge of the seat hit the sitter just behind the knees, which may cause an uncomfortable slump position in older people. (Young people never sit in chairs, anyhow, they sprawl.) Note that when you sit on a hard surface such as a bench, your forearms are only 2 or 3 inches above the bench surface. If you are planning high arm rests they had better not be too close to the hips. Do you want the back high enough for a head rest? Etc.

In other words, you had better take as much care in defining the concept of your chair before you build as you do in any product. Is it to be used to read in, lounge in, worship in, eat in, milk cows, wait a short time on, sit at a high bar, rock the baby in, fire a gun from, sleep in, wheel a patient around, or what?

After a few sketches of his settled-on concept, the young designer would probably be better off if he immediately built a small, rough, scale-model (say 1/4 inch = 1 inch) out of balsa wood, soda straws and pins, or what have you. This model can be rather crude, but it shows up design problems like rack, unsightly braces, negative spaces and silhouette form at a glance or push.   And now the details of joints and bracing can be much better understood and drawn more convincingly. The exact way material is put together is often the most nebulous part of a young designer's plans. How easy it is to join two pieces of plywood or metal by merely drawing them together on the paper with the magical pencil. But when the time comes to actually make the joint, "Aha, there's the rub." Mitre? Butt? Weld? Hinge? Corner brace? Glue? So this is one case where making the model BEFORE the plans helps the young designer understand better what his drawing details must contain.   This is step 2.

Sketch Models

Illustration board, Soda straws, pins, and paper can simulate

Mahogany board, PVC or Brass pipe, screws and Naugahyde.

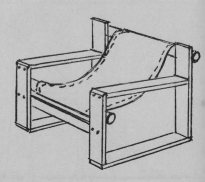

Step 3 would be for the designer to make a few detail drawings to show the instructor exactly how he proposes to fit the joints together. The drawings should be self-explanatory. A rough, sloppy sketch with lots of verbal blab to explain it is not professional. (I have made it a rule that detailing can not be accompanied by verbal explanations. If I can't understand it from the detail drawing then the designer obviously needs more training in drawing to communicate his ideas.) And it's "Back to the drawing board," or get out your notes from your perspective and rendering class.

Step 4 may be to make a client presentation of a front and side elevation with a three-quarter view of the entire chair on Canson paper or other ground, plus drawings on tracing paper of the parts. The tracings can then be made into blueprints or brown lines.

Or Step 4 can be skipped, and the student and instructor can discuss building a prototype.

Once the student has communicated his approach to the instructor satisfactorily he can go ahead and build the chair: Full Scale if he has the money and desire. Half Scale would make it the size of a child's playroom chair; Quarter Scale would make it the size of a doll's chair.

The student who knows materials and tools and has a sense of structure from previous carpentry or metalwork, such as those who have worked on hotrods or boats, could probably build a full-scale chair. Those who are just getting their feet wet in Product Design might be better off to take on a half-scale or even quarter-scale model. Also when facilities are limited and the classroom is only so-o-o big, there are space advantages in working on and storing the smaller models.

Quarter Size (Dolls)   Risse   Half Size (Children)   Casner   (PVC)

Willard

Paek

Jorjorian

Danley

Masse

How about the class designing chairs that are asymmetrical? That'll shake 'em out of a rut.

Full Size (Adults)

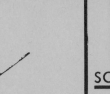

SOLAR COOKER

Use heavy, shiny aluminum foil plus chipboard, cardboard or other scrap materials to work out a collapsible solar cooker. A four-foot diameter circle or square is approximately large enough to provide enough heat for cooking.

The reflector may be somewhat parabolic in shape to insure that the parallel rays of light from the sun are angled correctly and sufficiently concentrated at a focal point where the cooking takes place. However, if it is designed to be perfectly parabolic, you'll have a temperature at the focal point that will drill a hole through your pans, so it can be dangerous. Experience has shown that a plain, spherical surface with a radius of 36 inches will do cooking more efficiently because the rays cannot be focused at a single point.

Auto headlamps are the reverse of the optics of a parabolic solar cooker. The light rays from the bulb located at the focus point are reflected out through the front of the lamp in parallel lines.

Several wedge-shaped flat sections of foil are usually easier to form than an entire spherical bowl surface. They also tend to spread the focused rays out to the size of the kettle instead of one focal point which might be dangerous. In a reflector that is spherical, with a radius of 36 inches, the spot of most heat concentration will be about one half the radius of the sphere, say 18 to 20 inches from the surface of the reflector. (See diagram.) An adjustable grill can be mounted as shown to hold kettles. Black utensils work best as they absorb heat.

REFLECTORS

GLASS

INSULATED OVEN WITH GLASS HOLDS HEAT.

OPTICS

Opaque Projector

Make a cardboard mockup of an inexpensive, home opaque projector. Keep it simple, with a minimum of adjustments. For example, focus might be accomplished by moving the entire projector toward or away from the screen, instead of using an adjustable lens. The inside of the case may be left shiny and bent in such a way that the sides act as reflectors to increase the strength of the light source, instead of adding reflectors later. (See **completed model**, page 149.)

Slide Projector

Make a similar model of a simple home slide projector. Insert and remove slides by hand. No complicated slides, or slide holders — just use a simple slot track integral with the case, etc. Can you make it simple enough so it can sell for $10 retail? Cost of lens would probably be determining factor.

IN

OUT

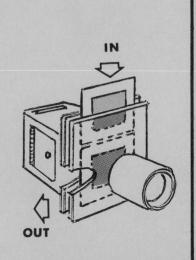

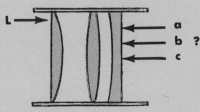

LIGHT RAY L WOULD COME OUT AT a b or c ?

EXHIBIT DESIGN

Traveling Display Panels

Design a set of display boards that can be joined together without separate fasteners or hardware of any kind. They must be modular units so the entire set can be packed in one rectangular box and shipped on to next destination. Consider hardness of panels. If presentations are pinned, taped, glued or matted on the panels, will that be significant in determining whether the panel should be made of tempered or untempered masonite, celotex, pressed wood, ply or laminates of vinyl and steel, or what? Make model to scale out of mat or illustration board.

Okay. Then try a set with fasteners. Separate hardware or crimped into panel edges? Rods, clamps, bolts, or slip fit? Sketch is sufficient.

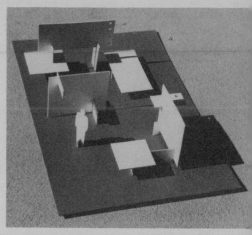

Rosenthal

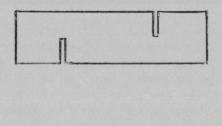

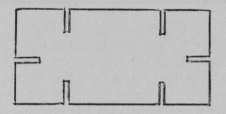

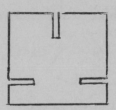

Point-of-Sale Piece

How about one single 3-D shape that can be unfolded to stand by itself as a Point-of-Sale piece? Six feet high. Triangulation for rigidity. Made from heavy cardboard, all to be printed one side only. (This means that the original flat sheet was sent through the printing press so that all type and art were printed on one side of the cardboard sheet only. This saves printing costs.) But you, the designer, have to make certain that, after unfolding and erecting, the printed images don't show up on the inside of the structure. Make model from box board 1/6 or 1/3 scale.

Double Tripod

Work up sketches for a back-to-back collapsible tripod which would support irregular forms like rocks and driftwood or spherical objects such as world map globes or beach balls. It could also be used to support a flat panel. Design 3 different sizes: one 4 feet high, one 2 feet high and one 1 foot high. Can you design these supports from cylindrical or square conduit with slip fittings so that when disassembled or collapsed the three pieces could fit into one container 6 inches square and only 3 or 4 feet long? Music stands are inexpensive approaches, whereas good camera tripods are an expensive approach. Can you simplify, or use existing hardware to make it it sturdy, good looking and reasonable in price? It is an item

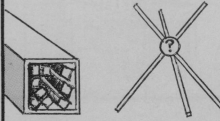

183

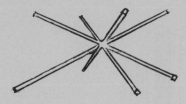

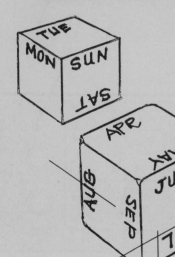

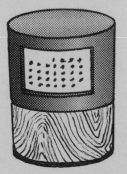

Note letters must be parallel to edges when on perspective drawing.

that might be an exhibit designer's multipurpose support for a variety of forms. It does not have to be chrome-finished either!

When you are certain you have it solved, make the middle-sized one and see if it really works. Don't fret about plating or finishes. Just see if you can make it work mechanically. Use electrical conduit for the tubing, if necessary, but see if your sketches stand up to function.

If each end had 4 legs (a quadruped), instead of 3, it could be laid on its side also, eh? — or (?)

TIME

Desk Calendar

Design a simple container to hold 4 cubes. Two of the cubes have the digits 0 through 9 arranged so that one number only shows on a side. These two cubes are side by side so the 31 days of the month can be shown from 01 up through 31 by turning the blocks each morning. The other two cubes are side by side and show the month and day of the week. If you want to add a section for clips or 3 x 5 note paper, etc., go ahead. The first cube has the numbers 0, 1, 2, 3, 4, and 5 on it. The second cube has the numbers 1, 2, 6, 7, 8, and 0 on it. (Design the type face so the numeral 6 reads as 9 upside down. In this way we can get all 31 combinations on 12 sides. Pretty clever, huh?)

The months can be placed two on a face. (2 times 6 = 12 months.)

The days of the week can be placed on the fourth cube. One on each of 5 sides and two days printed on the 6th side.

Permanent Calendar

Work out the following information so that it can be applied to a flat wall calendar with sliding-face opening; a cylindrical desk calendar with rotating inner cylinder; or a wristwatch strap calendar.

Two basic monthly calendars can be combined to form all possible combinations. All that is necessary, then, to use this calendar year after year, is a device that will expose the present month only.

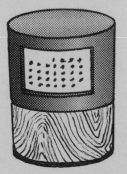

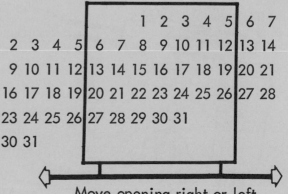

Move opening right or left.

184

PIANO TUNING INSTRUMENT

Piano tuners tune pianos by listening to the beats or vibrations of each note and then tightening or loosening the tension on each string or set of strings. The tuner has to have an ear for it, a built-in metronome, or he won't be able to work at this type of job. There are approximately eleven million pianos in the United States and not enough piano tuners. The younger generation has not become interested in it in late years so the average age of piano tuners today is about 58.

If you read the above article in a newspaper or journal would it occur to you that there might be a market here for a piano-tuning device that could be used by the piano owner whether he had an "ear" for it or not?

Could you design a tuning fork or rod or electronic metering device with some type of readout so that anyone could tune their own piano? Possibly a sensitive microphone plugged into a meter or oscilloscope would move a needle or show a wave pattern which could be matched with a previously set pattern for the correct harmonic beat or whatever it is called.

Those of you with Hi-Fi, radio, or telecommunications backgrounds might be able to get help from an engineer in that field or possibly a member of the Piano Technician's Guild, of which there are about 1800 members in the U.S. Even if the device proved to be expensive, it could be rented out by music stores.

Diagrams, circuits, illustrations, as well as precise operating instructions should be part of the design project.

The socket wrench which fits over the tuning pins in a piano is called a tuning hammer. Would applying the torque-wrench handle principle to this tuning hammer assist the amateur tuner in any way? Would knowing the exact amount of torque applied to each pin help him later to retune a string that had slipped?

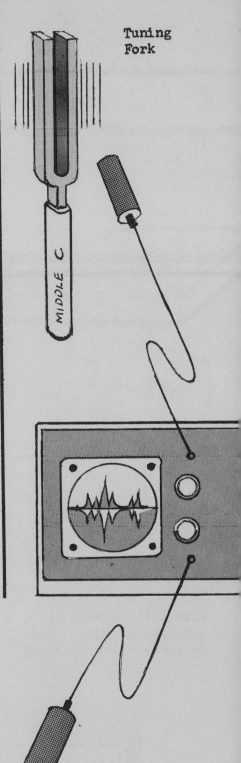

Tuning Fork

MIDDLE C

Middle C

Piano String

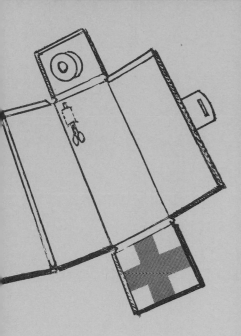

COMMUNITY

First-Aid Kit

Design a flop-open first-aid kit for firemen or paramedics:

Requirements:

1. Must be stiff enough to withstand reasonable crushing.
2. Case must withstand fifteen minutes in 280-degree heat for sterilization.
3. No metal or metal hardware to be used.
4. Must have handle or belt hook-on device.
5. Red cross must appear on three sides.
6. Must contain:
   1 roll gauze, 1 roll adhesive tape, surgical scissors, surgical knife, 6 gauze pads, cotton dispenser, 4 swab sticks, 4 tongue depressors, 1 unit alcohol or antiseptic, 1 hypodermic syringe and needle, 1 unit anaesthetic, 1 plastic tube for tourniquet or tracheotomy.

Park Fountain Model

Make a model of a park fountain using cardboard, doweling or welding rod, etc., to simulate:

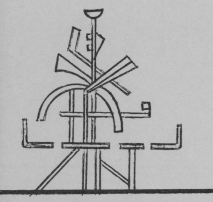

a. Concrete ?
b. Iron ?   (See concrete fountain, page 36.)
c. Aluminum ?

Classroom Chair & Desk

Combination school desk and chair for inexpensive classroom furniture.

Community Open-Air Center

Work up sketches for a small open-air stage and activity center in a small town. If you wish, carry the presentation into finished drawings.

Seating and stage	Lighting
Forum area	Storage
Lecture equipment	Toilets
Cinema, music, slides	Dressing rooms

Protection against:

Vandalism	Hail
Snow	Sun
Rain	Wind

Minor sport demonstrations:

Gymnastics	Dance
Badminton	Archery
Fencing	Etc.

ARE ALL CAST SHADOWS ON STAGE OK ?

STUDIO PROPS

In 1974 more than 60,000 people were victims of violent crimes in
Los Angeles County. The criminal today has all kinds of rights, even
up to free legal protection. But who pays the victim's hospital bills?
Who reimburses him for loss of wages or other damages? Does he get
free legal help also? A newscaster on a national network will be
making a special report on the plight of the victim in the aftermath of
violence, and will show that legislation protecting the rights of victims is sorely needed. A society that cannot prevent violent crime
should at least provide aid as well as legal assistance for the victims
as surely as it already does for the criminal. The newscaster has
asked your Design Group to make a series of from 3 to 5 three-
dimensional structures which will provide vivid graphics, either photographic or symbolic, of crimes of violence: Robbery, bombings,
beatings, sniping, or rape might be possibilities. Your structures
will be lighted dramatically and placed in the TV studio so the
camera can bring them into focus as different sections of the report
are narrated.

One simple way to start is to forget drawing and instead use a scissors
plus several sheets of 3 or 4-ply Strathmore or 6-ply poster board and
cut out a few quick models about 12 to 16 inches high. Forget realism
for the moment and merely cut out symbolic figures of a hulking
figure with pistol, bomb blast, screaming woman, figure with upraised
club, etc. Then notch, bend, interlock and build these quick thumb-
nails.

Just as one sketches rapidly to get his thoughts down on paper regarding an elevation or plan view of a product design — so, in this
case, cut, fold, notch and build rapidly to get the feel and composition of the structure BEFORE you settle on the exact shape of the
subject matter. When you need to design complicated three-dimensional forms that do not resemble the typical box shapes, elevations,
and plan views of so many man-made products, it is a good idea to
scissor your shapes first. Drawing is a very difficult way to envision
complex structures in the round unless you are an experienced
designer.

Full-figure paper sculptures are frustating.
This is one of the toughest problems in the
text, believe it or not. The student will
need triple the time he estimates, and most
will give up in exasperation.

Paper sculpture always looks so easy, yet
it needs to be practiced like any other
discipline before you become a Leo Monahan.

Single spot lighting can make these sculptures appear very dramatic.

REUSE RECLAIM RECYCLE

Milk Carton

Redesign the outside printing on a half-gallon plastic or paper milk carton so that it can become a plaything or toy instead of being thrown away immediately. It can be printed to resemble a milk truck, for example. Dotted lines can show the child where to cut or fold so that the carton resembles a dump truck, boat, monorail, etc. Could each side be a different building material, such as brick, concrete, wood siding, wallpaper, with doors and windows added here and there? Then the kids could save a bundle of them and erect whole houses. It could be printed to look like a shoe so when a hole was cut out above the heel the foot could be inserted and the clown could clop, clop around the house laughing insanely.

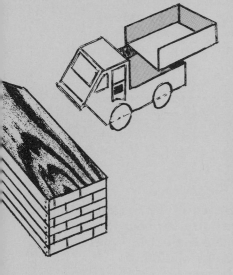

Maybe each side of the box could have a different western building; front of a saloon, stable, hotel, corral, etc., so they could be used to set up a small western town along with his Cowboy and Indian lead (Ooops — plastic) men. How about a train: half-gallon loco-motive, quart coal car, freight car, passenger car, pint size water tower or switch?

The dairy's name, date, kind of milk, etc. could be incorporated into the design, or possibly one side left blank for dairy information, such as the underside of the truck.

Any kind of design that increases the length of the use of products in our society helps the ecology. Containers, particularly, lend themselves to reuse if any intelligence at all is applied to their design and production.

Five-Piece Trash-Can Module

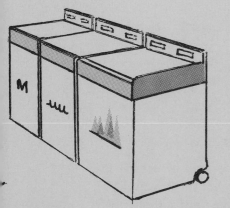

The warnings of scientists and ecologists for the last ten years are finally being substantiated. An energy shortage is really here. Gas rationing is being discussed by the president of the United States. Prices for food have been going up in many areas because of farm shortages. The herring base of the fish pyramid off the California coast has just about been fished out. Fishermen's boats are rotting in the docks and many fishermen have had to turn to other jobs. There may come a time when junkyards are an only source of metals and recycling becomes a must, not an option.

Let us assume that you have been retained by a client to design a plastic trash-can module of five units to be used by homeowners. The City recycling plant has specified five separations, so that re-claiming of trash can be conducted with a minimum of sorting. Wet garbage, burnables, glass, plastics, and metal are the rough classi-fications. The problem is to use color or form to enhance recognition. Hauling to and from curb, ease of emptying and cleaning, covers to keep out animals, and a base to minimize tipping are some other parameters of design.

Regarding Energy Use & Future Potential

According to a Rand Corporation study, "California's Electricity Quandary" of February 1973, technology offers no short- or medium-term solutions to the energy crises , so demand for power must be curtailed. The idea of using our energy more efficiently and conserving our resources is not a new idea, but Rand Corporation has lent considerable respectability to the idea of DECREASING demand instead of INCREASING supply.

A preliminary report of the Ford Foundation's Energy Policy Project, "Exploring Energy Choices," maintains that demand (for energy) can be stabilized and reduced. It goes on to say that reduction of demand would not necessarily cause lower standards of living. Section 7 of the report plugs for ZEG (Zero Energy Growth) instead of following our present trend of uncontrolled growth. ZEG could mean sane economic growth as well as allowing the less privileged to catch up. It is not possible within the scope of this text to list their reasons. But it should certainly be part of your education as a designer to look up these reports in your library and acquaint yourself with some of the practical and philosophical arguments presented by these "think-tank" groups.

Current Regulator

With the above in mind: Can you design a unit that could be attached to the incoming electrical line on every home that would limit the amount of electricity per hour that could be consumed by the home? Let us suppose a family could have 6 electric lights burning brightly, but as they turned on more and more lamps the bulbs would burn dimmer and dimmer until, eventually, if they left lights burning all over the house, it would be impossible to read easily in the study, and someone would have to run around and turn off the unused lights to get the rest back up to normal brightness. It's a cinch that many of you students have heard your dad & mother plead with you to turn off lights that you're not using. But being immature, and not having to pay the light bill, you fergit and go your own "free" way. Also, if you are running the toaster, a flat iron, and an electric drill at the same time, you may find you have to turn off one of the heating appliances to get that drill up to revs again, etc. Economic restriction of energy of this type certainly works no "hardship" on the people involved. It merely forces them to stop wasting our resources.

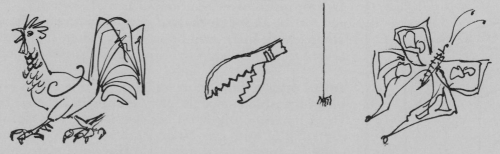

An interesting book on the Product Designer and his awesome effect on the ecology of the world is "Design for the Real World," by Victor Papanek, Pantheon, Random House, N.Y., 1971. A particularly neat chapter on "Bionics" shows how nature can provide the observant designer with a wealth of ideas regarding structures, force, communication, and motion.

WIND POWER

About 70% of the inhabited world has fairly steady prevailing winds of which the average wind velocity per year is about 10 miles per hour or better. This doesn't mean the wind blows every day, but it does give an indication of available power, because 8 to 10 mph winds can move a windmill, turbine or rotor. How about designing a windmill, wind rotor or wind turbine that would turn a small generator or an alternator which would feed electric current to a bank of storage batteries? From the batteries, then, an inverter could be used to transform the direct current (DC) of the batteries into alternating current (AC) which is needed for lighting, television sets and appliances.

Wind generators could be used by homes to save costs even though the homes were already tied into utility lines.

In "The Mother Earth News," March, 1974 issue, published in Hendersonville, N.C. 28739, Michael Hackleman points out in an article titled, "The Savonius Super Rotor," that, although the prevailing wind blows consistently from one direction, there are gusts from other directions. These gusts make up only 35% of wind time, but can provide 75% of the wind power if the wind machine is designed properly.

For example, when a gust comes from a different direction than the prevailing wind, it takes several seconds before the vaned mills or swivel turbines can turn into the gust. By the time they turn, the gust has often died away. But the rotor speeds up immediately from the gust and thus provides more energy. Rotors can take a wind from any direction at any time.

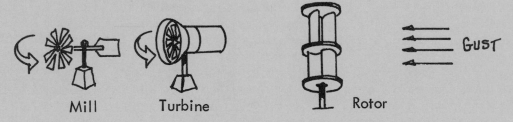

Mill　　　Turbine　　　Rotor　　　GUST

Rotors can be made by cutting oil drums in half vertically or forming half cylinders out of thin metal or wood. These half-round blades can then be mounted on a round base. Below are several possible variations of blade placement, looking down from the top of the rotor.

For testing these different arrangements of blades, cylindrical, cardboard oatmeal boxes can be cut in half vertically and mounted between round discs of heavy cardboard in the variations shown above. Then design a simple bearing mount and test each unit by fastening it on top of the bearing mount and place it a precise distance in front of an electric fan set at constant speed. Which configuration works best? How would you calculate which model rotor was whirling fastest?

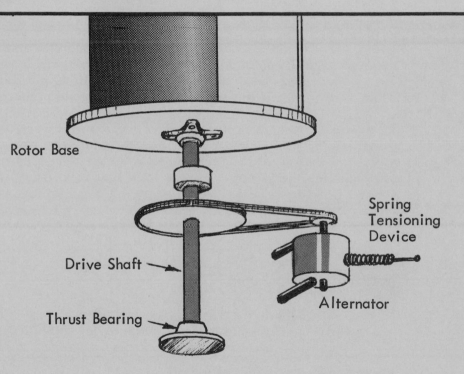

Rotor Base

Spring Tensioning Device

Drive Shaft

Alternator

Thrust Bearing

What design would be the most foolproof to transfer the power from the drive shaft to the alternators? Gears, chain drive, friction rollers? What ratio? That is, do you want the alternator to spin same speed as driver or twice or three times faster, or what? You'll have to talk to someone that knows alternator speeds and output and then figure how fast average wind speeds will drive a particular rotor or turbine. A rough estimate is that a 25 mph wind will turn a large rotor (say 9 feet high, 2 feet diameter) at 250 revolutions per minute with alternator going five times that. Alternators have different amperage outputs, such as 45, 60 and 100. Would your storage batteries be 12 volts?

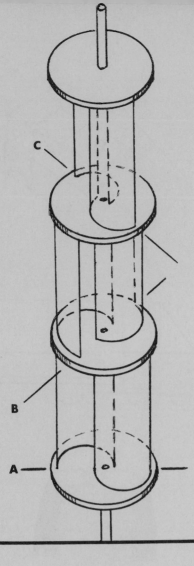

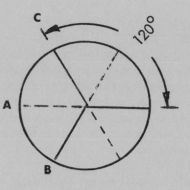

Set diameter line of
each rotor at 120°
from each other so at
least one blade will be
facing the wind regard-
less of gust direction.

What design would you devise to prevent damage to these wind machines in high winds? Dampening (braking) might be possible thru a
centrifugal governor that would come in only above certain rpm's. Or
how about adding more load, such as gearing in more alternators at
high speeds? The machine might be tilted, folded, or equipped with
gates to restrict air flow, or what?

Earthmind, a California farm research center at 26510 Josel Drive,
Saugus, California 91350, specializes in energy experiments with wind
and sun. They have some pamphlets available on how they designed,
built and compiled data on both wind and solar energizers. Federal government bureaus may have some information also.

SOLAR ENERGY

There are a variety of small Selenium or Silicon solar conversion units or
cells which harness sunshine and provide enough electricity to run small
motors. Edmund Scientific Company at 600 Edscorp Building, Barrington,
NJ 08007, has some cells available from $1.50 on up. Instructional
pamphlets on these solar cells and on solar heating for houses are also
available. Could you design a small toy boat with a propeller that
works off a solar energy cell on a sunshiny day? There is an interesting
article titled, "Solar Cells," on page 53 of the December, 1974 issue
of "Popular Science" magazine, which describes the manufacture of
solar cells and prophesizes their use in public utilities by 1990.

Four companies: McDonnell Douglas,
Huntington Beach, CA; Martin, Denver;
Honeywell, Mpls; Boeing, Seattle, have
been asked to submit designs for a
solar power plant by 1977. The ERDA
then intends to build one in the SW.

ENERGY-CRUNCH CAR

Design a small 3 or 4-wheeled car for, say, two people that could be used for driving a few miles to work or shopping. Go look at or write for information on the electric shopping carts that elderly people drive from the senior citizens center to the shopping mall. They are usually powered by storage batteries with an electric drive. The overall length is 6 to 8 feet with the wheels about 18 inches in diameter.

Keep it SIMPLE. And I mean SIMPLE. As an ex-Minnesotan I can very easily remember cars driven in sub zero weather without heaters. Do you really have to have backup lights? Cigarette lighters? Five ash trays? Double-speed windshield wipers? Buzzer to tell you when you're going over your speed setting? Adjustable seats? Come on! Rid yourself of your preconceived notions of car decoration and come-ons and get down to the real concept or reason a car was invented: to get you from one place to the other!

Yes, it was nice while it lasted (for those of us that could afford it) to be able to shave, take a bath, cook, and sleep in our cars while driving 75 miles per hour to a theatre 100 miles away. But with the energy crunch we will probably have to return to a saner mode of living. And whether the energy crisis continues or not . . . . we should nevertheless take a good look at our actions and stop wasting away our world on damn fool gadgets.

Can't you envision a dashboard, for example, that perhaps has no more than an ignition key and maybe a light switch on it? Why can't your windshield wipers be operated manually (as they were for years) or possibly by a foot pedal? If it's raining that hard that long, you shouldn't be out driving in it, anyhow. I'm talking about personal cars now, not the movement of freight. Stop being defensive and really see if you can't come up with the simplest car the world has ever known. Design it from your heart and forget the idea that a car has to be "loaded" in order to sell. We may be starting an era where a person's character and prestige are measured not by his movement and power but by the simplicity of his life. Maybe people have already started up that road. Big car sales have dropped, homeowners are turning off unnecessary lights, car pooling has started, the supersonic aircraft was lettered and tele- grammed to death, oil companies have been harassed until they have had to put on TV spots showing fishes can exist around tide- water oil rigs, etc. I have heard students talk about how they could easily get along with a bike, if they had to. At a party the other evening the couples were talking about how, "Maybe the family will do a little hiking or picnicking up the canyon occa- sionally now that we can't all drive 50 miles in different directions on Sunday " ; or even, "I think this energy crunch is one of the best things that ever hit this country!" All in all I sensed a strong undercurrent of humor and almost relief that we were finally going to come to our senses and take stock of our planet's inventory.

Saab Corp. in Scania, Sweden has been developing a steam car for the past 6 years or so.

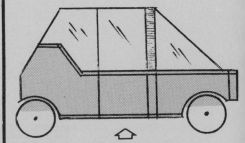

College of Art & Design Detroit, Mich., made a taxi so door would take wheelchair off curb.

So, Mr. and Ms. Designer, be aware that the world may be approaching a philosophy of simplicity and building for needs, instead of luxuries. I don't think it's a threat to our existence, or a time to frown and worry. I think it's one hell of an exciting challenge. Just think, as a designer you can now plead that decoration, tassels, luxury items are a threat to individual, national, and planet existence. You will not have to listen to clients choke out a grudging possibility that "Well, if you say so, I suppose it does look better. But I don't think the public will go for it unless we add those leather fringes, embossed buttons, a nail file holder, and disposable frying pan."

However, all kidding aside, our country is probably entering, in the 70's, a kind of twentieth-century rennaissance. There will be some tough decisions for people to make, shifts of both labor and management, and undoubtedly years of argument over national or state policies regarding our resources and energy. But in the long run we should be coming up with some answers that will not only conserve resources but will ensure we have enough work and take-home for everyone. The majority of our senators and representatives understand the possibilities, but they also need the active support of the public. You, the young designer, can certainly be part of this policy of conservation and better distribution. But you need to be aware of it and remain active during its expansion years.

This little car then may be a bit far out, but suppose it did work? In a sense it's kind of a symbol of a designer's philosophy. Could you design it so that everything on it could be reused to construct a replacement car, except the metal lost through wearing down of bearings, etc.? You certainly could take it through the presentation-to-client step. It might be interesting to compile its material and assembly costs, just to see how cheaply you could put out a "person mover." Could you use existing wheels or other parts? By sampling car wrecking yards or junkyards in your metropolitan area for small 18-inch diameter wheels, you might be able to estimate how many of these small wheels were available in the United States if you did decide to build a thousand cars. (Junkyards may be likened to mines. With judicious digging you find almost anything, with the added advantage of having it allready to go. Removing a bit of rust is a lot cheaper than smelting the steel, sand casting, heat treating and machining the finished part.) Can you use braking energy to help recharge batteries? Can you power it any other way than electrically? How about foot pedals for two people with a geared flywheel and clutch?

Use your design ability to make it appealing esthetically too. Remember that the product itself is the best form of advertising. If your brainchild is simple and beautiful to look at, and is REALLY known for its economy, dependability, and sensibility, it will undoubtedly replace the big, bulbous, garish, grinning-grilled, recessed-headlamped, shag-carpeted, gas hogs that choke the roads at present.

In five minutes, how many items can the class list that are unnecessary in a typical medium-priced car?

HOUSE TRAILER BUSINESS VENTURE

To sleep two. Frame dimensions. Axle & wheel design. Trailer hitch. Springs. Butane? Refrigeration. Screen door. Fenestration (window placement). Stove. Toilet? Sink. Storage. From diagrams to sketches. From "Things to Include" list to specifications. A final Canson paper presentation to scale, mounted on mat board to include:

a. Plan view
b. Elevation of inside cross section side view
c. Elevation of outside side view
d. Elevation of inside cross section front view
e. Elevation of outside front or back view
f. Several interesting detail drawings
g. Typed or well-lettered explanations plus possible legend for scale, etc.

What are your costs? List your design time. Estimate material costs and labor to build. Might be interesting, after you have done this, to seek an interview with a trailer manufacturer. Take your presentation along with your cost estimates and see what he thinks of your efforts. Gird up your loins a bit and be prepared for a variety of interviews all the way from, "Humph, boy do you seem stupid!" receptions to the more usual one of, "Gee, those are neat presentations." "I feel you're on the right track with your cost estimates, although I think you underestimated labor by about 20%." "Here, lemme show you some of my figures on a job we had last month and their relation to material costs." "You know, I've got a son, 14, who is getting real interested in design. Who did you say your instructor was?" "It's almost 12 o'clock. How about having lunch with me?"

While you are estimating your costs for probably the first time, be aware that the cost of designing and building one trailer (or other product) is somewhat deceiving. A young businessman starting a production line for selling in a competitive market is often fooled by the costs on the first trailer he builds himself. He is very apt to think, "Boy, I built that trailer for $450 worth of materials. How can they possibly charge $2,500 for the same damn thing at "Sportways?" "Come on, Bill, lets start building these babies, sell 'em, and make a mint." Months go by and although the first few trailers are sold at a lower price, the boys begin to discover they don't seem to be making much profit.

A variety of economic dragons start rearing their ugly heads:

a. City zoning laws prevent them from producing trailers in their own garage. They have to rent an old shack on the edge of the business district.

b. Utility bills seem to climb higher as electric welders, furnace operation in cold weather, increased lighting costs from night work, and extra trash collection add to the "Outgoing" column under operating expenses.

c. Small businesses have to report income to Bureau of Internal Revenue. These reports must be supported by records. Bill kept the records the first few months on a few sheets of paper. As customers ordered, requested deadlines, cancelled, paid part, or were late in paying, Bill found he could not keep it up and still find time to sleep and work on the trailers too. Result: Hired Mary as part-time secretary. Three on payroll now, instead of two.

d. A small fire one Sunday night destroyed a corner of the building and most of a partly built trailer. All told, they were lucky to get off with about $1,000 damage. No one had thought of fire insurance. So now they talk to an insurance agent and take out a modest policy for fire, theft, etc., and throw one in for liability too. Premiums will be paid yearly, but they have to start putting 1/12 of it away every month to be certain they have it to pay come January 15. Aha, now they have found out what accruals are, six years after that business math teacher explained it to them in high school.

e. Mary called the Internal Revenue Bureau to ask a question regarding sales taxes and social security reporting. The man helping her on the phone happened to mention depreciation along with some other items on checking account records, until Mary realized she was going to need more help. After calling on a friend who was majoring in accounting, they decided they were going to have to hire an accountant part-time, also, or really get fouled up. Four on payroll now, instead of three.

f. Their biggest order in ten months of operation was for six trailers. They missed their deadline commitment to this client by three weeks, because a trucking strike held up an expected delivery of 24 wheels. After this they decided to stockpile essential materials so an emergency inventory would enable them to keep manufacturing for one month, in spite of shortages. Where to store it? A warehouse rental (more Rent Outgo . . . "Damn.") And more money tied up in that month's supply of steel, wheels, plywood, welding rod, extruded aluminum, etc. (The money invested in material inventory does not earn interest in the bank anymore either.) Who is going to run over to the warehouse and make certain our stock is up? Is it insured? "Well, no. You see that's in a different building." However, the warehouse manager will provide you insurance for a slight increase in the month's rental. Is this larger or smaller than what you'd pay for insurance from your first agent? Who is going to find that out?

I think, by now, the student designer is beginning to get the picture of what real costs are. I do not want to discourage young people from starting a business. In fact, running your own business is a very complex creative game that takes plenty of guts. America is one place where you can pull it off. And even if you don't make it, it is no disgrace. But the young designer should understand some of the problems young companies run into and not have a smug attitude or know-it-all opinion on how to run a company. Try it once, first, before you sound off too loudly on how to lower costs. Taking one or two classes in accounting and business law contracts will often be the best insurance for keeping your own design office solvent or at the very least giving you empathy with production foremen or corporation presidents.

In one company I worked for, that employed about 45 people, it was figured out that it took one man's full-time output merely to stay even with government forms, edicts, and regulations. This meant that if the company was to not only survive, but take advantage of every legal opportunity allowed them in figuring costs, assets, liabilities, tax deductions and resultant profits, this guy had to spend almost eight hours per day reading government regulations on everything from transportation, licenses, tax seals, depreciation, capital gain, to dates on perishables, as well as keeping up with law changes, bills before congress, and import or export tax variations. This man had to be intelligent with a modest background of law as well as economics. His salary was $18,000 per year. This is not huge by some corporation standards, but just remember that you have to sell one * hell of a lot of products or design time to make $18,000 over costs just to pay this one man's salary. It ain't easy.

So when you are figuring costs in these textbook design problems, realize that you are listing primarily design time, materials and labor only. This is good practice for the student designer to know what costs are involved with the immediate product. But these costs should not be confused with the other total overhead costs of running a solvent business. The two boys in our trailer success story discovered the hard way why that "same trailer at Sportways" sold for $2,500. So do remember that, although the designer is part of the input, the final decision of setting a competitive sales price and retaining a margin of profit is determined by many other factors and is primarily the responsibility of management.

---

* For more information regarding government restrictions, read "Putting the Cuffs on Capitalism" by W. Guzzardi, Jr., an article on page 104 of the April, 1975, issue of "Fortune" magazine — another warning to us that deepening penetration of government into every business decision is breaking down our market system of free enterprise, initiative and risk taking. Continued far enough, this type of government control will result in a boring, sterile society which will have no incentive to invent, design, or take risks of any kind. Witness the low national product in communistic and socialistic countries. Take away the individual's chance to make a profit, and you also take away incentive.

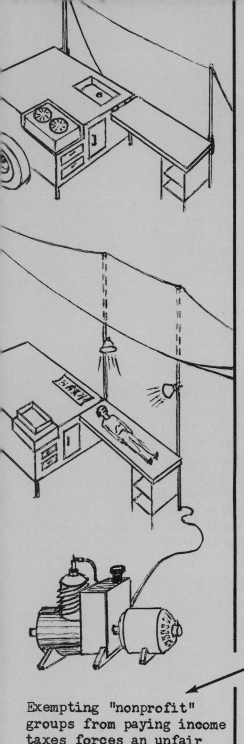

Box Trailer Kitchen

Plan this stowaway kitchen to be semi-expandable. Table tops with folding legs could be pulled out to form a "U" plan at camp site. Holes or slots might accept long telescoping poles which support a canvas sun shade or lean-to side drops for wind protection. Corner legs drop down to keep trailer level when detached from car? Water supply tank with spigot or pump. Outside hose bib would allow filling from garden hose? Fold out rack to hold roll of paper towels. Sections for stove, pots, cutlery, dishes, soap, garbage, tools, cutting block, food, ice box drainage, etc.

First-Aid Transport

Could you adapt your trailer design so it could become an emergency first-aid station or "hospital" for underdeveloped areas or rural communities? That is, when an epidemic or disaster hit a place that was separated from doctors, medicine and hospital equipment, several units of this type might be hauled in by horse or jeep. Items in the trailer might include an expandable roof, anaesthesia equipment, operating table, electric generator, fuel, lights, hot water supply, medicines, surgical instruments, disinfector, cases for fragile instruments such as microscope or hypodermics, etc. A talk with a doctor or nurse (particularly those who have seen service near a front-line hospital during a war) would give you a good idea of the absolute minimum basic supplies necessary to help sick or injured people during disasters.

And don't kid yourself that all areas in the world can support central hospitals or even first-aid stations, much less helicopters to fly in medical aid. If you have traveled much around the world or even in your own country and kept your eyes open, you will undoubtedly realize there are many local areas that could be helped out of poverty and despair by the simplest type of modern technology. So don't "pooh pooh" design that deals with basic, or even primitive principles, as being unnecessary. The world can use billions of dollars to put spaceships in orbit, but it still allows millions of people to starve to death each year. There is plenty of design work yet to be done at the gut level, which doesn't require a Ph.D. in physics to accomplish. All it takes is a student like yourself, a good education and a client. The world MAY be approaching the time where "Foreign Aid" will really mean aid instead of armaments. If so, you young designers of the next generation may find recipients of foreign aid, state governments, United Nations, UNICEF, or even local communities among your most appreciative clients. And by that I mean you will be finally well paid for your services, instead of having to donate them as a charity gesture, which happens too often today when "nonprofit" groups ask designers and other artists to donate their time for free.

Much of the above, obviously, is dependent on whether the threat of war is diminished. IF we can solve that; IF we can hold down the population explosion; the solution would only be a matter of time. So, hang in there, and keep solving problems at all levels. The world needs you.

SCHEDULING

Scheduling a project, whether it be designing for a deadline, making a prototype, or setting up an efficient flow chart for a production line, is a very necessary operation in many instances.

Usually two factors are considered to avoid bottlenecks. The job to be done or part to be made is one factor. And the time estimated to do it is the other factor. When these two factors are estimated and charted along a main time line, they show clearly exactly what stage the product is in. Otherwise the information is often buried in a stack of papers on the manager's desk, and heaven help the production department if he ever gets sick or dies.

In the chart below, the circled numbers indicate the start of a component which is to be assembled with either another part or in the main unit. The arrow point then indicates the end of the time alloted to produce that particular component. The heavy line through the middle of the chart might be likened to the assembly line in an automobile plant. The frame starts at the left by being welded together from incoming structural steel members. As it progresses along the line it is joined by axles, springs, steering mechanism. Then the body is lowered in place after it has been welded together. The engine is dropped and bolted in, doors are fitted, drive shaft inserted, etc., until the car is started and driven off the end of the assembly line to be inspected, tested and delivered. The above is a rough example, but the principle involved is the same in a design office. Some things cannot be accomplished until other parts have been decided upon or components made to specifications are completed.

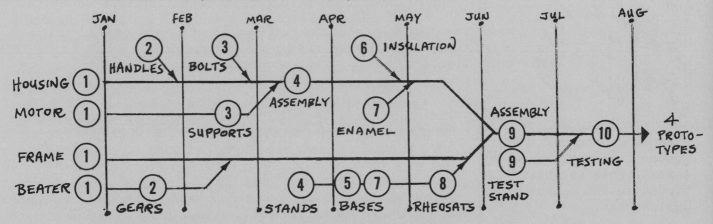

If design is not scheduled efficiently, bottlenecks occur, and time is lost. Time is money, and when costs go up, clients become irritated, to say the least. Allow for contingencies and don't accept unrealistic deadlines from a client, "just to get the account." Always remember, if you do fail to deliver on the deadline in the contract, the client will let everybody in the world know about it. Most top designers won't accept rush deadlines. And usually after a few weeks the client comes back to them, hat in hand, because he couldn't find anyone else to accept it either. Clients in a big rush are clients who are disorganized. Beware.

TEAM PROJECTS

During the second quarter or semester a team project often proves quite illuminating to students. They learn that they get outvoted, they sometimes have to take on responsibilities designated to another member of the committee because that person caught the flu, or maybe just isn't coming through, etc.

For the first team effort it is often better to redesign a package than to develop a whole product. The time span is shorter. The students have fewer problems to resolve. They are usually better acquainted with the esthetics of flat surfaces, etc.

Packaging

Divide the class into groups of 4 or 5 students each. Do this by merely following the alphabetical sequence on the roll sheet. First four names are team one. Second four names are team 2, etc. Each group then chooses their own mode of operation. They may appoint a chairman, they may not. They may designate one member as the research gatherer, one member as the renderer, or all may take a whack at it and then vote on the best. Instructors I have talked with about this seem to prefer this method. (When the instructor selects the top 6 students in the class, appoints them chairmen, and then assigns each two medium and one "poor" student as a team he has immediately put his stamp of hierarchy on the team. The members are ipso facto junior assistants to the chairman. Enthusiasm ebbs immediately, and too often each chairman has the first and last word without much argument from the rest of the team, because the instructor has shown by his choice exactly what he thinks of them.)

Each student then brings in 4 different packages from the supermarket or corner grocery or drug store or toy shop or elsewhere that he thinks need redesigning. Each team then has to select ONE package out of the 16 that they think they can best handle.

For research they must contact the local sales representative of that product and get as much information as they can regarding the design and production of that package. Tell him tactfully that they are a team of students who has as an assignment the packaging of his product and, if possible, would like to arrange an interview with him. The team may decide to send only one or two members to the interview, but that is up to the team and the interviewer. The package must meet legal requirements. In other words the team cannot delete all the fine print and use only color and one or two main heads. Laws require, in many cases, that the ingredients be spelled out on the front of the package, etc. These things should be learned from the interview also. In some cases the sales rep or managers contacted have become so involved with the students that they have asked to come to the classroom on the day the team presents its package to the class just to see how the presentation comes off.

Thanks again John Caldwell

The teams agree on a specific deadline date. The grade assigned to the presentation and package is the grade each member of that team receives. If some members don't come through on their part of the project and the presentation folds, then all members receive the olde F. If only two members of the team do all the work and it turns out great and the presentation gets an A, then the other two members of the team also get A's. This should be fully understood at the very beginning of the project.

The teams can use any method they want for building the finished package. The package can be built from scratch by scoring and folding box board. (See box problem, Section 3.) Old packages can be sprayed with paint, or pasted over with color-coated paper. Any photo-mechanical process like 3M Color-Key or brownline process can be used to cover the models also. Transfer type or wax backed color film are possibilities. In other words, the best way to get the job done is the way to go, depending on the experience and expertise of the team.

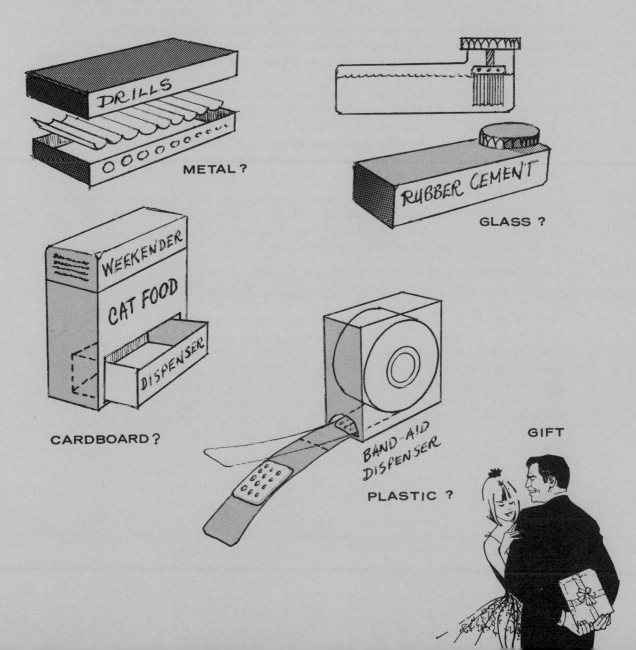

METAL?

GLASS ?

CARDBOARD?

PLASTIC ?

GIFT

Radio

Design a radio that would cost about $2.00. This again could be a team problem. But in this case, the students are forced into an area of design that usually is completely new to them. The typical student will rear back and say, "But I don't know anything about a radio... I can't even change a light bulb.. eh,eh." The theory and circuits of radios are really not that complicated, and many 12 year olds have built a simple set as far back as 1915. So let's go.

Three students per team. Can you use an ear plug from a hearing aid; can you use an old piece of screen for an aerial; a piece of wire and a nail to stick in the ground for a ground; an oatmeal box or cardboard orange juice can to wrap some thin copper wire around; pick up some old crystals or transistors; dig in junkyards for some copper wire; etc.? Find a ham radio operator, or the radio instructor at your school, or dig the circuit diagram out of the library. If there are many broadcasting stations in your area, how will you channel your set ? If there is only one station you are in luck.

Do you think you could find a market for this $2.00 radio set in poverty areas, instead of just considering it a gimmick? Would it have any application, for example, in a large forest or swamp area where children have no money to spend and no central school to attend, but where a central radio station might be able to broadcast lessons at certain times each day? How many of these sets could three of you make in four hours?

Radio Circuit

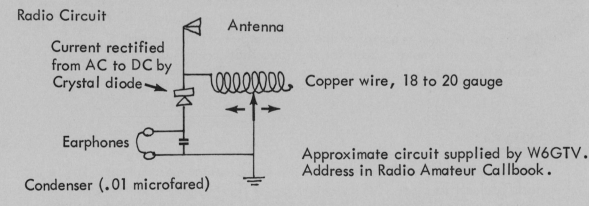

Antenna

Current rectified from AC to DC by Crystal diode →

Copper wire, 18 to 20 gauge

Earphones

Approximate circuit supplied by W6GTV. Address in Radio Amateur Callbook.

Condenser (.01 microfared)

Paper Clothing

Do team research on various types of paper. Don't forget that there is paper manufactured for industrial toweling, packing, and bags, as well as the more common kinds used for stationery and printing. Select a few kinds that are rather tough, as well as flexible and somewhat water resistant. Set yourselves some other specifications that you think are necessary in order to make the material adaptable to your method of manufacture: are you sewing, crimping, or gluing the seams together? Now actually make a jacket or a dress and keep track of the time involved, once you know what you're doing. Figure your costs and present the project to the class as a possibility of low cost, throwaway clothing which is also biodegradable.

FUNCTION & FORM

Does form really follow function?

"Form follows function" was a phrase often used during the early 1900's to describe a so-called principle of esthetics. Yet this statement has been challenged in recent years. The argument against total acceptance of this platitude is that there are many, many differently formed devices in the world that perform identical functions. Airplanes, for example, vary in form from bailing wire, sticks and linen concoctions to the slick tube of a commercial air liner. Yet they all fly. Now one can say, "Well, the bailing wire provided tension to prevent the wings from sagging just as the perforated magnesium-aluminum cantilevered structures support the wings in the air liner. They each are true to their form." Well... maybe,   but that is really stretching it. Nature, herself, provides animals and plants with a million different forms and they all perform functions that are almost identical in many cases. They walk, run, jump, dig, swim and fly. They see, breathe, ingest, exhaust, reproduce, hear, make noises, and feel. Nature is a top designer when it comes to producing things that work, and the forms of life are infinite. Give a function to three different designers working in isolation and you'll probably end up with three different forms, each of which performs the function. So let's not get too hung up on the "form follows function" deal.

Could you write a rebuttal of the above argument? In other words, can you make a case for the idea that form DOES follow function?

RECALL

Seeing and imagining seem to use the same mental mechanisms. If you ask yourself to close your eyes and count the number of doors in your house, you will probably discover you will move imaginatively through your house counting each door as you come to it much as you would have had to do if you were really moving and seeing the doors in a real house that you just bought. Although these mental images are constructed from stored information, they can be almost as vivid as the actual visual perception and have the added attributes of fantastic speed of travel or speed of execution. Could you design a test to determine what persons have the most rapid recall of these imaginative images? Let the class take the test and see if there is any correlation between speed of recall and the more creative designers in the class. Check with one of your psychology teachers who is interested in psychometry.

SUGGESTIONS FOR DISCUSSION OR SHORT REPORTS

1. Can plastics be made biodegradable?

   If plastic trash was ground up as fine as sand or dirt, could it be mixed in with hard soils such as adobe clay to keep the clay from packing? Are the chemicals in the plastic too toxic for plants? Are they released that fast? Plants need minerals. Are there minerals or chemicals in some plastics that could be released or combined with other chemicals that could provide food for plant life?

   An article in the "Northwestern Engineer", March, 1974, on Biodegradable Plastic Films told of a chemical engineer, Michael Gradassi, who developed a type of biodegradable film. He used an Amylose film, which is water soluble and falls apart when wet, combined with other chemicals which made the film more stable and then introduced glycerol as a plasticizer. The glycerol was used because it is readily assimilated by fungus and thus helps the film to "rot" when it is thrown away instead of building up or polluting the natural environment.

2. There is an old saying that a strand of a spider's web is stronger than a strand of steel of the same diameter. Can you prove or disprove this? Through expert opinion? Researched reports? Or experimentation? What does stronger mean: harder to cut? Takes more tension or torque before it breaks? Bends at a larger angle before it cracks? (See Properties of Materials, page 58.)

3. Modern veneers, or laminates, simulate many materials such as walnut, ash, tile, marble, brick, etc. Does this bother you esthetically? Do you subscribe to covering steel with plastic to make it look like wood? Is your acceptance of these materials based only on economics? (That is, "Well, who can afford walnut or mahogany anymore?") If a laminate simulates wood so perfectly that the senses cannot detect a difference — is that honest design? Or are we so far along that any report made on this subject would be merely beating a dead horse?

4. Is the giant jet air liner good design?

   Some aircraft manufacturers and air lines are subsidized (given money) by the government to keep them solvent. Why doesn't the government also subsidize a small manufacturing plant or your design studio when you approach bankruptcy? Does the noise, smog-exhaust, low scrap value of the short-lived, multimillion-dollar aircraft warrant its use by people who think they must go somewhere in a hurry?   Should manufacturers be completely indifferent to reuse, scrap separation or recycling of our natural resources in regard to such large items of public property as a billion dollars worth of air liners? Of course this is the type of design problem you cannot do much about — yet. But you must be aware of the larger over-all problems too, or you may well become one of the "If it's bigger, it's progress " advocates. What suggestions would YOU offer to a federal senate committee investigating subsidies?

5. Designing for the Future

Invent or modify a product for production four or five years from now. How would you use the time lag to make the consumer ready for the product?

This happens quite often in spacecraft hardware. Companies put out artist sketches, mocked-up models or retouched photographs with typical interfaces plus estimated specifications as to size, weight, etc., of components which will not be ready for two or three years. Thus the prime contractor, who is designing the larger unit, will be able to take into consideration the possibilities of design that are not now produceable but will be available several years from now when the spacecraft nears completion and "the state of the arts" will have advanced.

6. Depreciation & Obsolescence

Assume you are head of a successful design firm that specializes in small transportation items such as, motorcycles, bicycles, outboard motors, and electric shopping carts. A client has asked for your opinion on a possible new policy.

The client would like to call in each item sold, at specific intervals (say time or mileage) and change worn out parts on a schedule. These repairs and replacements then would supposedly prevent the product from wearing out but would have to be accomplished at exact intervals or the warrantee would be nullified. The client would like you to recommend which of the above four products might lend itself to such a plan. He would, of course, want your pros and cons on the whole idea with some type of conclusion.

If you do decide it would be feasible on at least one of the items, write the warranty in clear, concise language so it is understandable by the customers buying the product. Will replacement parts & labor be included in original price or paid for as used? How long will warranty last? What about rising prices over a long period of years? If your client goes out of business, who takes over warranties? Who decides if customer has abused product just to get new part? Etc.

7. Rhythm

Is the essence of rhythm the recurrence of relationships, not necessarily units? Must the intervals be "expected" by the audience? Walking, running, swinging, jumping are fun partly because they result in recovery from a fall? The majority of people do not like interruptions of their daily rhythm or routine of living? There is time rhythm in music (ear), space rhythm in art (eye), and kinesthetic rhythm in sports (muscular). Are there any products that develop pleasurable rhythms in taste? In smell? Can you think of any possibilities? For a toy? For a cook? For a winery?

8.   What three questions would you ask an applicant to help you evaluate his creative thinking?

How would you answer the three questions yourself?

9.   You are a designer for a plastic-forming company. The management feels that coming legislation in regard to waste collection and recycling trash may bring a demand for special containers of all types. The headings below exemplify various areas which might require separate design approaches, and you are asked to sketch a few ideas regarding containers or plastic products that would help control the collection, and sorting of garbage and trash.

1.   Reducing flow into waste stream from industry
2.   Waste collection or disposal
3.   Recycling or reconditioning plants
4.   Household garbage and trash
5.   Transportation wastes such as crates, cartons, tires, etc.

Select a single unique item that you feel has the best chance to sell within your local area. Develop it for presentation.

10. You are manager or Senior Designer of ONE of the following offices:

a.   Industrial Design Associates
b.   Structural & Expandable Metals Co.
c.   Architectual Ceramics
d.   Outdoor Furniture, Inc.
e.   Wrought Iron Furniture Co.
f.   Toy Company
g.   Parks & Playground Equipment House
h.   Sporting Goods Mfg. Co.
i.   Outdoor Lighting Co.
j.   Advertising Agency
k.   Specialists in Exhibit Display
l.   Product Testing Division of
      Better Business Bureau
m.   Consumers Research Magazine
n.   UNESCO Design Section
o.   UNICEF Design Section

List 10 specific things you would like to see in the portfolio of a young designer who is applying for a job in your particular area of design.

11. Think up 6 original games, toys, or things to do for a boy or girl, 7 to 10 years old, if they had to stay indoors on a cold rainy day. From these ideas, sketch up one design, package it, and estimate its manufacturing cost, manufacturer's selling price, and the retail store's selling price.

12. You are working as a junior designer for a company that manufactures electric lighting fixtures. In view of the coming energy crisis, the design head has asked you to sketch up a variety of household lighting fixtures which use kerosene, butane gas, or other fuels. Such lights already exist as campers' lanterns but are not particularly suited for a hanging chandelier over the dining room table, or an entrance hall light fixture. Mantles or flames must be enclosed, fuel tanks must be easily changed or filled, and lighting the lamp must be reasonably simple with controls for brightness.

13. Can you form a block that will fit exactly when inserted into each of the following holes? Make a perspective drawing of it.

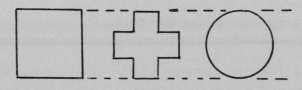

14. Can you form a block that will fit exactly into each of the following holes? Make a perspective drawing of it.

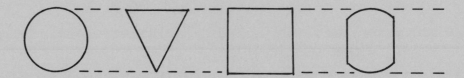

15. Following are four draftsman's views of a small component case: front and back elevations, top and bottom views. The other two opposite side views are not shown. Can you visualize what shape he is drawing? Make a perspective drawing of it.

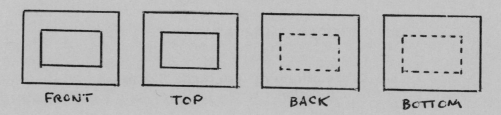

FRONT   TOP   BACK   BOTTOM

16. Tradition

"Know it... Don't worship it." Explain this statement with examples.

17.   From the following set of facts try to determine:

A.   Who drinks water?
B.   Who owns the goat?

1.   There are five houses.
2.   The Englishman lives in the red house and owns a horse.
3.   The Spaniard owns a dog.
4.   Coffee is drunk in the green house.
5.   The Ukrainian drinks tea.
6.   The green house is immediately to the right of the
        ivory house.
7.   The Hindu owns a pet snail.
8.   The Protestant lives in the yellow house.
9.   Milk is drunk in the middle house.
10.   The Norwegian lives in the first house.
11.   The Catholic lives next door to the man with the fox.
12.   The Protestant lives in the house next to the house with
        the snail.
13.   The Jew drinks orange juice.
14.   The Japanese is a Buddhist.
15.   The Norwegian lives next door to the blue house.
16.   In each house there is only one nationality, one pet,
        one religion and one liquid drink.

18.   Government in Design

Government cannot legislate good design. But it can help by hiring good designers and reputable architects for its buildings and similar design areas. It can promote an atmosphere in which cultural activities may flourish. A National Council on the Arts has been established. If you were asked to speak before this council, what type of policy would you advise them to follow? How could they encourage state governments to follow similar policies? Ditto, city or local governments? Would you suggest sponsorship of any particular legislation? What specific things could they do to better the look of cities? Absentee landlords, for example, allow buildings in the center of cities to become fire-traps and infested hovels, because they don't want to spend the money to fix them up. Do you think increased taxation on these kinds of neglected properties would encourage the owners to keep them in shape?

Write an outline (not a report) of your ideas and present it to the class for their evaluation. If you come up with some reasonable good ideas, why not send a letter to your government representative in that area, signed by class members who are interested, urging some action on these problem areas?

19.   Eliminating Competition

Can you think of 2 products for each of the following reasons that have had extraordinary success in sales volume:

Quick entry into market (e.g., Hula-Hoop)
Lowcost (e.g., Ball Point Pen)
New patent
Secret formula or process (e.g., Coca-Cola)
Foresaw fashion trend

20.   Community Improvement

Select a small town in your area and suggest how the community might improve itself by the use of local materials and local talent without spending a lot of money. One large project or several smaller ones might fill the bill.

21.   Reduce Trash Mail

What suggestion do you have to cut down the waste of trash mail and the indiscriminate use of our mail system by firms that mail a "shotgun" blast of unwanted flyers to thousands of people who promptly toss them in the wastebasket and thence into the city waste system?

22.   Reduction of Hiway Litter

What would you incorporate into automobile design to help stop highway litter, which costs the U.S. government $500,000,000 per year to clean up?

Do you think this is a problem worth working on? Well? Is this product NEEDED? Or is the real requirement an improvement in human character?

23.   Throwaways

People today are educated to impermanence. That is, articles used in an industrialized society are often meant to be thrown away. It is only in primitive or poor societies where functional everyday items are treasured. Even architects realize they will probably live to see their buildings torn down. "If It Works - It's Obsolete" is a needle-point tapestry that hangs right alongside "Home Sweet Home" today.

What factors in the world do you think might change this attitude in people ?

If it does change, and people start looking again toward permanence, do you think they will be more sensitive to esthetics? That is, if the consumer knows he has to use an item for 25 years, will he be a little more sensitive to how it looks? And how will this attitude affect your own philosophy or the policies of the design firm you work for?

24. Contemporary Design

Would you say that the main visual difference between architecture today and that of pre-1900 period is that we do not use ornament, sculpture, or symbols of structure such as pilasters, columns and capitals to decorate our buildings? What other factors characterize contemporary design of all products?

25. Environment

Is man alone in that he is the only living creature that changes the environment to suit himself instead of adapting himself to the environment?

26. Know Yourself

Some people live art but don't DO art.
Some live art and DO art also.
Some DO art and don't live it.

If you subscribe to the above statements, how would you describe the characteristics or actions of each personality? How would you evaluate their attitudes?

For example, a well-known local designer and teacher leaves trash all over his campsite. He also allows his students to leave the design lab and power-tool room in a shambles for the next class to clean up. He may DO art, but he sure as hell doesn't LIVE it!

27. Polarized Glass

Polarized glass may be likened to a venetian blind. The glass is similar to a lattice. Let's say the light that comes through a piece of this glass comes through in thin parallel bands between the slits. (See diagram A.) Now if a second piece of polarized glass is placed over the first, with slits parallel to the original, the light keeps coming right on through both pieces. (See diagram B.) But if the second glass is given a 90°-turn, the criss-crossed slits cut the amount of light coming through. (See diagram C.)

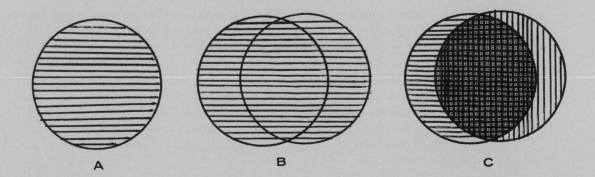

A   B   C

Can you envision all cars having polarized glass headlight lenses with vertical lattices and all cars having polarized windshields with horizontal lattices? Would headlight glare at night be cut to a minimum? However, when light leaves the headlights through the vertical slits and bounces off the roadway, grass, rocks, trees or signposts, it is reflected in many directions and theoretically loses most of its polarization. Therefore the driver would be able to see objects in the light of his own headlights without any noticeable loss of brightness.

Can you mock up an experiment and present it to the class to test the above theory? If the test is inconclusive, how many other possible applications can you envision for polarized glass?

28.   Optical Glass

The latest dark glasses on the market turn dark when exposed to bright sunlight and then become clear when the wearer goes inside where the light is less intense. Do you think you could use this information to produce a reuseable negative? For example, printing plates in industry are made from negatives. These negatives are a thin plastic film with a light-sensitive emulsion on it that turns dark and hard when exposed to light, developed, fixed and washed. Thus, light images are recorded on the film as dark areas, and the dark areas of the subject become the transparent areas on the negative. After the negative is used to make the positive printing plate the negative is thrown away. The silver from the wash water and that left on the negative is thus lost forever. Tons of silver are used up every year by the photographic and printing industry, so it won't be long before silver will be in short supply.

Let us suppose you could use a flat sheet of this "photogray" glass for a negative instead of the regular film negative. This glass neg is then placed in the copy camera and exposed to the light rays reflected from the art work. It becomes darkest in those areas where the light from the art work is brightest and remains transparent in those areas exposed to the dark areas of the art work. The intermediate grays are resolved into the intermediate values on the glass negative.

Now, quickly, before the glass has a chance to start clearing itself, place the glass neg over the light-sensitive emulsion on the metal printing plate and squirt a burst of white light thru the glass neg onto the printing plate. You now have a positive image transferred to the plate.

The glass negative can now be stored and used later on another job, as it will soon clear itself and revert to its original condition. Boy, think of all the silver you have saved.

Crude perhaps, but this is a typical way design ideas are born. Stay curious about everything, seek relationships, and remember principles.

There are other conditions to be met when copying intermediate values (halftones) in art work. Ordinarily, the film negative in the copy camera is covered with a contact-screen film which separates the light rays into a series of tiny dots. For a more complete explanation of this process refer to chapters 12, 13, & 14 of William Bockus's "Advertising Graphics," second edition, 1974, Macmillan Publishing Co., Inc., 866 Third Ave., NY 10022.

Could you talk with an optician or get some information regarding this peculiar type of glass and conduct a demonstration in class with some fast photographic printing paper, for example, instead of a printing plate, just to test the feasibility of the idea? And let me warn you that almost everyone you seek information from will want to know what you're doing, and when you tell them, they will point out to you ten reasons why it won't work.

Maeterlink : "Each progressive spirit is opposed by a thousand men appointed to guard the past." Or somewhat more up to date, "Behind every successful man is a woman telling him he's wrong." An old engineering saw goes something like this, "Illigitimi non carborundum." Translation: "Don't let the bastards wear you down."

EYEGLASSES   (PINHOLE SPECTACLES)

The pinhole camera needs no glass lens. The small opening allows only the minimum number of light rays to enter that are necessary to define the image on the film. Focusing is unnecessary. If a similar hole is drilled in an opaque material and placed in position in front of the eye, it acts in the same way to define a more precise image on the retina in the eye. In other words, the pinhole reduces aberration or distortion by insuring a clearer image CLOSER to the eye.

However, the eyeball swivels in different directions, so the pinholes must be drilled in various positions on the spectacle "lens" (see diagram). You will have to experiment with other hole configurations, sizes and spacing. Holes drilled too close together may cause double vision, for example. If you would like any more information, Diana Deimel, owner of Diana's Nutrition Center, 505 S. Glendora Ave., Glendora, CA 91740, has experimented with the production and marketing of this type of spectacle.

Do you think there might be a market for this type of inexpensive spectacle in the underdeveloped areas around the world? Prepare a marketing analysis in which you propose steps to lessen or overcome resistance from opticians, optometrists, oculists, medical authorities, eyeglass manufacturers, The Pure Food & Drug Administration, or even people's vanity.

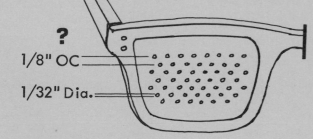

?
1/8" OC
1/32" Dia.

Try crosses or alternate slits with holes. Left-to-right eye movement is more critical than up-and-down movement.

Ophthalmologist ?

RETINA PLANE PERCEPTION

As you design, in your drawings or constructing a model, try to delete the subject from your visual concept. Sounds ridiculous, but designers often have the capacity to see the three-dimensional form as sets of two-dimensional relationships of color, value, shape, size, and position. (See sketches.)

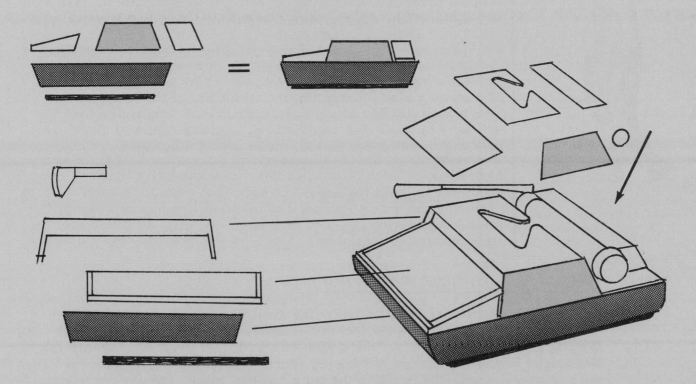

The artist's ability to perceive these patterns in flat plane is somewhat unlike the nonartist's perception. It is a kind of discipline or order of the above elements imposed "on" or "in" the 3-D form. It has an almost subliminal effect, and its fulfillment is usually recognized in the first split second of observation. "I like that" is often the immediate reaction of the consumer.

There are several magazines concerned with contemporary design that might be included on the lab bookshelf:

"Industrial Design" One Astor Plaza, 1515 Broadway, NY 10036
"Domus" Via Monte di Pieta 15, 20121 Milano, Italy
"Abitare" Via Guerrazzi 1, 20145 Milano, Italy
"Mobilia" 3070 Snekkersten, Denmark

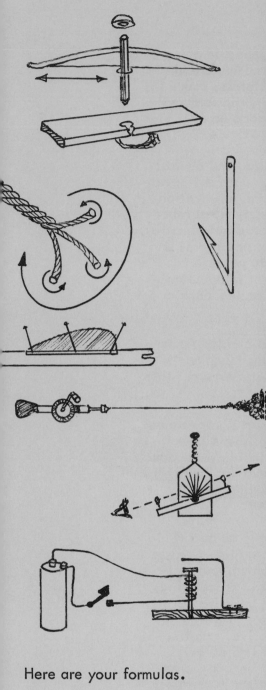

BASIC PRINCIPLES

Around 50,000 to 10,000 years ago primitive man began designing tools and a variety of objects. In a pinch, without using any modern manufactured item such as matches, magnifying glass, hammer or knife could you:

1. Make a flint knife?
2. Make a fire?
3. Find clay somewhere and make a pot that would hold water without crumbling?
4. Skin an animal and then treat the hide so that it would not become brittle and hard?
5. Make a drill?
6. Make a needle?
7. Make a fishhook and line?
8. Make a bow, bowstring and arrow?

Or using anything you can find in the room, could you:

1. Make thread from the chunk of raw cotton or wool lying on the instructor's desk?
2. Make a loom and describe its operation?
3. Make a small sheet of paper?
4. Make ink?
5. Make a sundial?
6. Make a sextant or crude equivalent? Take a reading on the North Star at night and report to the class the latitude of your position as indicated by the reading.
7. Make a telegraph sender key and receiver? Use the Morse code or make up one yourselves and send a message across the room.
8. Demonstrate to the class how a telephone works?

A bonus of an extra "A" in the roll book to the person who can teach the class most quickly and clearly the principle and function of the spool-needle and bobbin-shuttle mechanism of the sewing machine. No one may have more than 3 minutes to make his presentation. Clear, readable diagrams usually win. Heaven help the instructor in his evaluation; and don't be surprised if nobody can explain it satisfactorily.

And if you want to hop 10,000 years into the future when YOUR culture may be referred to as "those twentieth-century primitives" read "Intelligent Life in the Universe" by Shklovskii and Sagan, 1966, Holden-Day, Inc., San Francisco, California, USA. A Russian and an American have written the book together. In one example, they discuss variations of the Dyson Sphere: That is, building a 10-foot thick wall of material from the planet Jupiter into a shell or sphere to encompass the sun just outside the orbit of the earth. This, you see, would then trap all of the sun's rays, from which we could use all the sun's energy instead of harnessing only an infinitely small part of that force at the present. How'd you like to work on a design project like that?

An interesting book on way-out physics is "Roots of Coincidence" By Koestler.

Here are your formulas.

Volume of a Sphere $= \frac{4}{3} (3.1416)R^3$

Surface of a Sphere $= 4 (3.1416)R^2$

(Where R = Radius of Sphere.)

Diameter of Jupiter is 87,000 miles.
Diameter of orbit of Earth is 186,000,000 miles.

Is there enuf matter in Jupiter to build such a shell?

214

ROBOTS

One of the ultimate designs will probably be a robot, capable of a variety of actions, that responds to human instructions. We already have many robot-like devices: They wash dishes or clothes, make toast, scrub & wax floors, dig wells, move tons of dirt, dig ditches, fly aircraft for hours on end, and fabricate complex parts on a factory production line.

Cybernetics (sī'ber netiks) is the study of communication and control which are common to living organisms and machines. With the development of memory banks, information retrieval, and miniaturization, it is becoming possible to program (wire up) a reasonably sized robot so that when given a STIMULUS through its eyes (photoelectric cells), its ears (telephone receivers) or touch (pressure plates or buttons) it records and categorizes the various SENSATIONS into patterns of PERCEPTION. It then "thinks" by retrieving pertinent information from its memory banks and RESPONDS or acts upon this feedback or "experience."

As we all know, there is a tremendous difference between the human brain, which stores (learns) information by itself, and the machine computer, which stores (learns) only when fed a magnetic tape or other device.

At any rate, cybernetics is a fascinating study requiring a background in both biology and engineering. About 1938 an English mathematician suggested that a machine could be said to be thinking if it could carry on a conversation with a human being in another room and that the human could not tell whether it was conversing with a machine or not. The conversation could be by sound or readout.

As an overall robot design philosophy, it would probably be best if robots were used to complement human weaknesses. In other words, use the machine's strengths, such as endless repetition without fatigue, infra red sight, millions of responses per second in calculating, operating in extreme heat or cold or without oxygen, unlimited strength or power, etc., as a reason for their existence, rather than try to replace the human being. A human needs creative, constructive work, EVEN IF HE ISN'T QUITE AS GOOD AS THE ROBOT. A human being without reason for existence is a very unhappy creature. The century may not be upon us yet . . . but time is certainly getting closer when the designer will have to ask himself: "Is this robot necessary?"

An interesting novel regarding robots and the future is Jack Williamson's "The Humanoids," Lancer Books, NY. The most recent publication was 1963, I believe. It's a chilling picture of robots, which have a built-in trait of protecting humans, doing everything for people and allowing them absolutely no risks at all.

Another book, "I, Robot", Issac Asimov, about 1950, Signet Books, NY., is a series of short stories concerning technical and philosophical problems of handling advanced robots.

Below are two columns headed HUMAN and MACHINE which list
some relationships or contrasts between the two. As more and more
differences are resolved, we come closer and closer to ourselves.

	HUMAN	MACHINE
1.	Can speak on any subject.	Can speak on programmed subjects only.
2.	Can hear language and interpret.	Can hear language but can interpret preprogrammed ideas only.
3.	Can touch and react to variety of stimuli.	Can touch, but reacts only to preprogrammed responses.
4.	Can write.	Can print out.
5.	Can play music.	Can play music.
6.	Can compose music.	Cannot compose music?
7.	Can dance, play tennis, etc.	Can dance crudely and can project images of action in three dimensions soon.
8.	Emotions affect responses.	Has no emotions.
9.	Gets diseases.	Unaffected by germs.
10.	Doesn't rust.	Rusts.
11.	Operates only in limited temperature range.	Can operate in greater temperature ranges.
12.	Can be overloaded with input.	Can be overloaded with input.
13.	Gets tired.	Does not get tired.
14.	Breaks down when parts wear out.	Breaks down when parts wear out.
15.	Parts replaced by cell growth or transplant.	Parts replaced by transplant.
16.	Makes mistakes when operating in perfect health.	Does not make mistakes when equipment is in top shape.
17.	Gets better with repetition.	Never gets better with repetition.
18.	Bad habits are reinforced by repetition.	Has no bad habits. (Poor thing.)
19.	Works slowly.	Works much faster than human.
20.	Can learn.	Cannot learn?

HUMAN	MACHINE
21. Has trouble retrieving information.	Has no trouble retrieving information.
22. Tends to forget.	Never forgets.
23. Can theorize regarding academic concepts.	Cannot theorize on anything.
24. Can make probability judgments .	Can make probability judgments .
25. Solves mathematical problems.	Solves mathematical problems.
26. Can define problem.	Cannot define problem.
27. Plays chess expertly.	Plays chess inexpertly.
28. Trial and error.	Set routine. (Are errors necessary for learning?)
29. Skeptical.	Dogmatic.
30. Can time actions roughly.	Can time actions precisely.
31. Open-loop circuitry. (That is, information from outside environment affects actions constantly.)	Closed-loop circuitry. (That is, information stimuli outside the system do not affect operation.)
32. Attempts to predict outcome.	Outcome set ahead of time.
33. Difficult to predict human behavior.	Completely predictable.
34. Can interpret symbols.	Can interpret symbols.
35. Frequent uncommon behavior.	No uncommon behavior.
36. Gives different response to similar stimuli.	Gives same response to similar stimuli.
37. Brain probably consists of sensory paths connected in parallel.	Brain usually consists of connections in series? Limited by binary (2-way) connections?
38. Outputs from memory banks are compared at various levels and passed on or cancelled out. (Yes, no, maybe, if.)	Outputs from memory banks are judged in limited fashion only. (Yes, no.)
39. Great number of recalls gives higher probability of unique or creative response.	Limited number of recalls restrict "creativity" of response.

HUMAN	MACHINE
40. Purposeful behavior most of the time. (However, even the behavior of an insane person may be purposeful to him. Shall we say then, purpose as defined by normal people?)	Purposeful behavior most of the time. (Unless a flywheel breaks.)
41. Recognizes patterns and makes complex correlations.	Recognizes limited patterns and makes limited correlations.
42. Has experience.	Has feedback?
43. Initiative can come from within.	Excitation must come from without.
44. Has self-control, inhibitions.	Has governors, valves and relays.
45. Can reproduce self.	Can only reproduce other machines now.
46. Very flexible. Can adapt to novel situations.	Inflexible. Cannot adapt to novel situations.
47. Makes decisions and acts upon insufficient evidence.	Can only make decision and act when has all evidence.
	We can put it in and we can read it out. Is the problem how to connect the two through judgment and logic?

To put it another way: Our inventions of the past have been to help man's legs (transportation), to help his hands (tools and appliances and machinery), to help his sensors (telephone, TV, radio). The inventions of the future will probably include helping his brain and nervous system. Computers are a start. And when you get a few breakthroughs in memory storage, retrieval, and analysis (judgment), then you will have the makings of a real robot.

Robot Umpire

Work out a system of light beams and photoelectric cells to register
the position of the baseball when it is thrown through the strike zone
above home plate. Let the position register on a readout panel located on the scoreboard. (LASCAR, Light Activated Silicon Controlled Rectifiers, or some such mouth-filling title, are the little
gadgets that hold a relay switch closed or open until the light beam
is interrupted. Burglar alarm systems sometime employ these devices.)

A bat swinging through light grid would of course make a "mark" on
the readout panel and would be read as a strike. Batters would not
be allowed to swing through the light grid once the pitcher started
his windup. This would keep the panel clear until either the ball
went through the light grid for a "called" strike or the bat swung
through.

Readout Panel

Solenoids or relays activate circuits
which turn on the corresponding row
of lights on the readout panel corresponding to the beams of light interrupted in the light grid. Where the
vertical band of light crossed the horizontal band of light on the panel
would indicate the position of the
baseball as it passed through the
light grid in front of home plate.

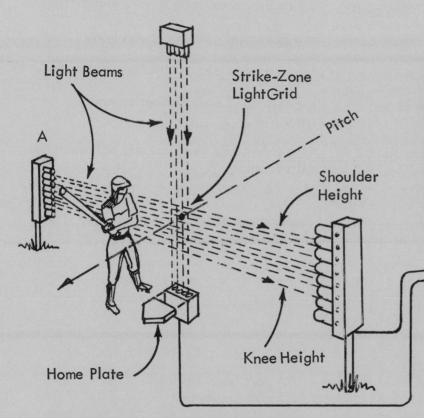

Light Beams

Strike-Zone
Light Grid

Pitch

A

Shoulder
Height

Home Plate

Knee Height

Photoelectric cells are arranged in a
line along the front edge of, or slightly in front of, home plate. Space
them about 1 1/2 inches apart so baseball could not pass between light
beams without interrupting them.

A similar row of small, focused, high-
intensity lights are suspended directly
above.

Photoelectric cells are arranged vertically on a post (? feet to the right
of home plate) from a height of about
1 foot up to a height of 6 feet. This
should cover the distance from shoulder to knee of all batters. Each cell
on the post has an individual switch
which can activate or deactivate the
cell. The batter stands in front of
the post, and the umpire deactivates
all cells above the shoulders and below the knees of the batter just before the batter steps into the box by
home plate.

A similar row of small, focused, high-
intensity lights are placed directly
opposite, about ? feet away from
home plate, so that the beams pass
over the front side of home plate.
(See post A.)

A simpler, cheaper system could be devised in which a light would come on if the ball passed through the strike zone or remained off if it did not. Although in this system it would be necessary to provide a unit that would not activate the light unless BOTH vertical and horizontal light beams were interrupted. Otherwise a wide ball outside the vertical beams would activate the light as it passed through the horizontal beams.

Consider the following: Are you replacing the umpire? Or are you only taking advantage of this robot's ability to "see" better? Will you pursue this design reasoning of better efficiency until the umpire is no longer needed? Does your product take all the "fun" out of the game by eliminating half the spats between batters, catchers, and umpires? Eventually, of course, you would have the game played by robots who never made a mistake, and the game score would either be 0 to 0 or 1000 to 1000. Are you real certain you want to design this product? Humorous now, but nevertheless you must learn to sit in judgment on your own design decisions. If you don't consider the future impact of what you do, who will? You certainly mold the physical environment around each of us more than doctors, lawyers, truckers, managers, and politicians. They distribute, arrange, modify, and plan, . . . but you actually create and CHANGE the environment. Think first. Is this robot necessary?

THE LAB ENVIRONMENT

It may be impossible to "teach" creativity. On the other hand we can assume that everyone has creative ability already built in, or the human race would never have gotten where it is. If one of our main objectives is to encourage creative thought processes, what are some specifics that might be applied to a product design lab situation?

One approach might be to break down our analysis into four sections: 1. Preliminary Preparation 2. Selection of a Problem 3. Solutions and 4. Evaluation of the Solution.

1.   Preliminary Preparation

TRAINING teaches you what others have learned.
Education teaches you what to do with curiosity.

There is probably a minimum of creative activity in a training program. The teacher may use creative ways to put across his material, but the learning of techniques is quite often a long-term process of practice, practice and more practice. The problems are usually rigid, given from authority, and the pressure to conform to certain standards is top priority.

Self-initiation is a definite part of creative activity. As a student you may be more interested in developing certain techniques than others. Can you set your own objectives?

a.   Do you want to develop craftsmanship in at least one area ? Two? What would they be? Sketching, rendering, matted presentations, working drawings, model making?

b.   Do you want to work with a variety of materials, or would you prefer to explore maybe two materials in greater depth? What materials: paper, markers, colored pencils, pastels, ink, paint, plastic, wood, metal?

c.   If you are new to the field of product design would you prefer that the instructor gave you options to choose from? Or would you rather have him direct you in set steps?

d.   Are you prepared for a certain amount of drudgery and boredom in learning what others have discovered about techniques and communication? Will a long-term goal (four years, for example, instead of a few weeks) help you get through these "practicing technique" periods? Are you aware that during the period you are sharpening your manual "tools" that you will also be tackling design problems that require thinking solutions?

2.   Selection of a Problem

In this second step we will assume we have already acquired a few techniques for communication. We can knock out reasonable free-hand sketches; we aren't too bad at perspective renderings; we know a few things about plastic, metal and wood; we can handle drafting tools and know an elevation from a plan view; with some help from the instructor we might even come up with a cross-section drawing. Now:

a.   Do you want to select your own problem? Or would you rather work from options presented by the instructor? Psychologists claim that self-initiation is indispensable to induce creative responses. How do you reconcile this statement with the hard fact that many of your early years in business or industry will be spent doing ASSIGNMENTS? If you can't pick your own problem do you feel your self-initiative has been stamped on?

b.   Suppose you are allowed to select your own problem and afterwards discover it has already been solved? Or that it wasn't much of a challenge? Or the instructor didn't think much of it? Or it was unsolvable? Are these conclusions bad?

3.   Solutions

Ray Bradbury said once, "I don't turn my back on my mediocre experiences. They are part and parcel of my creativity." But students often bemoan the fact that they, "... have to take all those dumb classes that don't have anything to do with art (or science or whatever)." Yet most creative people agree that in the long run ALL learning helps thinking; that saturating oneself with all kinds of information, relevant or irrelevant to design, helps creative activity.

a.   Do most creative solutions begin with a question? If you ask the instructor for answers concerning solution do you want him to answer immediately? Do questions settled by citing known facts from an "authority" hinder thinking, restrict discussion, prevent exploration and lead to a dead, dependent mind? Suppose the instructor refused to answer any question regarding solution of the problem? Would you feel he was lazy, arrogant, or that he was forcing you to THINK? Do you want the instructor to answer your questions regarding SOURCES of research information, or should you do your own digging?

b.   Do you realize that every time you are asked to solve a design problem you have to come up with a UNIQUE answer; a solution no one has thought of before? Do you realize that in many other classes you have to come up with an answer that is set beforehand? Identifying artists, buildings, or pieces of art in Art History is one such example. Dates, wars, economic trends, chemical reactions or rock formations are the subject of objective tests which can be graded "right" or "wrong" by a machine in many cases. In math the answers are in the back of the book, and if you DON'T come up with that answer you are "wrong." In a sense then, too many courses are training courses which have very little to do with sharpening your thinking apparatus. Product design, on the other hand, is one of the classes where creative thought processes are encouraged early in life.

Now creative thinking is often generated when something goes wrong or doesn't work right. Or it can be started by problems that stem from confused, unsettled, poor, or ugly environments. Therefore the solution IS a tough one, and the artist-designer has to learn to live with uncertainties and frustrations. You will undoubtedly be subjected to "failure" much more often than those who can "look up" the answer. But why not count yourself lucky that you're in with a creative think group where mistakes, wrong alleys, and change orders are a part of the ball game. Actually they are your bread and butter .., eh? And, boy, is the result rewarding! Artists all agree on one thing at least, and that is the immense personal satisfaction they get from producing an original work. It may not be the "best"; it may be "greasy, ugly and covered with hair"; but, by Abraham, I did it "all by myself."

So take heart, young designer, Don't be afraid to try everything and work out the bugs as you go. Remember that even if you DON'T find the optimum solution, you have nevertheless stimulated your synapses and practiced patterns of thought which gradually etch themselves on your brain and thus make creative thought that much easier the next time.

4. Evaluation

The skills were learned, the problem selected, the solution made; the plans were drawn, the presentation and models are finished. How do we evaluate? Well, before you can evaluate anything you have to know what your objective was.

Was your objective to:

Self-initiate research and learning?
Encourage original thought processes?
Make professional working drawings?
Lay out a top-notch presentation?
Build a craftsmanship-like model?

Will you need several different evaluations instead of one grade on the whole kit and kaboodle?

a.   Are you satisfied to receive a grade on your project and let it go at that? Would self-evaluation in several different areas (such as those above) help to point out strengths and weaknesses? Suppose you wrote out a short evaluation of your PROJECT from the standpoint of uniqueness, practicality and craftsmanship. Then wrote out an evaluation of YOURSELF regarding self-initiative, scheduling, work habits, and creative thought processes. Would you want to rank yourself against other students? Did you schedule with efficiency, or did you find yourself doing most of the work the night before it was due? Etc. Suppose you kept this written evaluation confidential until AFTER the instructor had critted your problem to see what points he made that were similar to yours or where you differed.

b.   Did you feel incompetent or worry about technical errors as you tried to communicate your ideas? Do you feel some further rigid, noncreative training problems in technique might be necessary now in certain areas? Which areas?

c.   Seeing that grades are almost inevitable,   if you receive a grade lower than the one you thought you deserved will this cause you to drop your major? Aw, come on. Are you learning? The statistics of attrition in a typical public college are something like the following: First-year Freshmen 12,000; Sophomores 6,000; Juniors 3,000; Seniors 1,500; and then 800 graduates. In the main, the dropouts are caused by the lack of drive. May I remind you that the Model-T Ford with a full tank of gas can outrace any Porsche or Ferrari ever designed without a gas tank. If you love designing, if you like to create ... then grit your teeth, hang on, stay tough, man the ramparts and keep your flags flying. The world needs you.

There is a very clear concise article titled, "Aids to Creative Teaching," by Dr. Ralph J. Hallman, in the May, 1964, issue of the CTA (California Teachers Assoc.) Journal. The ideas in this article are particularly important for any young designer wishing to promote his own creative behavior.

White to play, and mate in three moves.

10 GENERAL INFO.

TRENDS IN PRODUCT DESIGN

Below are listed some possible trends in the design field. These are, of course, opinions. It is difficult to forecast any important changes in marketing areas, but it might be to your advantage to read over the items and keep these possibilities in mind. Later then, as you apply for a job, run your own office or design products you may be able to spot significant trends that would aid you in foreseeing market changes or even protect you against a catastrophic bankruptcy. The all-around designer doesn't only design products for today; he also keeps his head up and his eyes open regarding the society he works in.

Shortages in metals and probably in plastics will very likely be with us for some time. Petroleum derivatives such as solvents and synthetic fabrics will probably be short also.

Cotton goods and wool may be harder to get. As other countries get more affluent, the demand for cloth rises, and as synthetics slow down, the crunch comes. Demand exceeds supply.

Expect some growing interest in design for the handicapped. The development of bionic motors may open new doors here. Related to this is growing research in medical areas for measurement instrumentation and devices to control diseases. Although "pure" research is the environment at most universities, there seems to be some possibility that commercial interests and objectives of a "practical" nature may become a greater part of the ball game, particularly in life sciences.

Possible growth of drive-in super-supermarkets. Redesigned warehouses may stock bulk commodities as case-lot pickups for the drive-thru customer. More emphasis on packing-case design, or bulk-item sacks?

Laws requiring pollution controls on industry have forced some plants out of business. A number of paper mills, for example, have not been able to spend the money necessary to control the wastes that were previously dumped into the rivers and thus are out of business. Paper will undoubtedly go up in price, and color selection will probably be limited because colors come from oxides, which cause the pollution and greater expense.

However, as a result of tougher pollution laws, there may be even greater expansion of recreation areas. This may happen not only on the national level but on the local level also, where small streams, parks and small wooded areas can be put into public use again. The smokestacks pouring smoke into the valley's woods have stopped; the fish in the stream can breathe again, now that the solvents are gone. What designs might be in that type of atmosphere again if the automobile is banned from these areas?

Expect a great increase in communications equipment and information retrieval. Keep up on laser developments, microfilming processes, cybernetics switching, structural foams, laminates and three-dimensional projection. Getting in on the ground floor of new developments is good insurance against obsolescence of your ideas. This does NOT mean to drop everything you're doing and plunge into the new invention area. That would be ridiculous. It does mean though that you could use a bit of your manpower and a bit of your resources to keep your group informed and experimenting in design areas within your plant capability.

Development of reusable or returnable items will increase. Conservation of energy is in — dissipation is out. I will wager to say that most of you students feel a little guilty every time you throw a food, pop, or beer can into the trash. And you should. That metal is gone forever. Recycling is a must, not an option. Be the firm that gets in on the ground floor of recyclable items.

City governments may turn more often to design firms for consulting advice on signs, malls, mini-parks, etc. The horror of rat-infested ghettos and the neglect of absentee landlords in the center of cities has started a backlash of disgust with public authority that will undoubtedly lead to more and more demand to clean up our cities. Does your office have communication with the city board of directors or planning commissions? Does your city have an "in-plant" division of design where you might find a niche? Would increasing taxes on rundown buildings help?

Corporations and even smaller businesses are becoming interested in community image, education and ecology. Safe-driving advertising by oil companies, park recommendations by savings & loan associations, employee recreation, helping clean up streams near an industrial plant have all become part of the present climate of business interest. As a designer you have to have faith in the fact that this interest in the ecology and fellow man is a deep, basic movement.

Undoubtedly some of the advertising about corporate interest in ecology is baloney and merely a gimmick to sell more products, or to relieve consumer fears without any real backup truths. But still the die is cast. What men (or corporations) start saying they are doing, is often followed by actual action. More legislation will probably be needed and is definitely forthcoming at the present time to force large industry to comply with ecological principles. Consumer groups are banding together here and there to investigate a variety of practices from safety regulations for atomic energy plants and the control of radioactive residue from "breeder" plants down to type sizes on drug cartons.

Ralph Nader, for example, has established two centers for investigation: One for consumer complaints and one for corporate employee complaints.

Public Citizen, Inc.
P.O. Box 19404
Washington, DC 20036

Clearinghouse for Professional Responsibility
P.O. Box 486
Washington, DC 20044

The world-wide repugnance of war as a method of solving anything is a start. Communication systems bring examples of starvation, neglect, and hopelessness into every man's home. As a result the generation that is now managing or entering into business, education or government is a generation truly dedicated to morality and ethics in a manner probably never seen in the world before. Certainly we are far from perfect, but within my lifetime, and particularly since the end of World War II, there have been significant changes in both government and business attitudes. Contrast society today in its efforts to help mankind with the hordes of Genghis Khan, and Alexander; the Roman Legions; Hitler's murdering of 6 million Jews; and Stalin's purges of 12 million Russians. We may just have something, you know, after all these years, and it behooves you, as a designer, to use your ability to make the world a better place by thinking before you design something. Is this necessary? Am I conserving resources? What is the corporation I'm working for really doing? Is it ethical? Because if you don't, who will?

			Prefix	Symbol
1 000 000 000 000	=	10^{12}	tera	T
1 000 000 000	=	10^{9}	giga	G
1 000 000	=	10^{6}	mega	M
1 000	=	10^{3}	kilo	K
100	=	10^{2}	hecto	h
10	=	10	deka	da
.1	=	10^{-1}	deci	d
.01	=	10^{-2}	centi	c
.001	=	10^{-3}	milli	m
.000 001	=	10^{-6}	micro	u
.000 000 001	=	10^{-9}	nano	n
.000 000 000 001	=	10^{-12}	pico	p
.000 000 000 000 001	=	10^{-15}	femto	f
.000 000 000 000 000 001	=	10^{-18}	atto	a
.000 000 000 000 000 000 001		10^{-21}	beep*	b

*One of the smallest units ever conceived. It has stemmed from the interval between when the signal light turns green and the gal behind you beeps.

Altho you will not be using many of the mathematical terms, you should be acquainted with the prefix and symbols in the middle range. When dimensions or other measures are specified in terms of milli, micro, mega, macro, or deci and deka, the designer should not misinterpret and translate tenths as tens or pull some other boner.

DESIGN OBJECTIVES IN MASS MARKETS

Mr. Del Coates, chairman, Industrial Design Department, College of Art and Design, Detroit, Michigan, presented a paper at the Eighteenth Congress of Applied Psychology, Montreal, Canada, in 1974, concerning "The Role of Design in the Preparation of Industrial Products." He presented an interesting idea regarding design and the mass market.

He assumed, in general, two main bodies of consumers: The traditionalist, who tends to buy more conservative design and comprises the greater percentage of the market; and the avant-garde type, who tends to buy a product that is more unique or "way out" in design and comprises a much smaller percentage. Coates's observations support the theory that if a designer designs a consumer product in the intermediate area that leans slightly toward the avant-garde customer, he will find a definite market therein, and after a period of time this somewhat unique design will have been accepted by the traditionalist market, and they, in turn, will then start to buy. Now if the designer designs only for the traditionalist market, he will probably experience similar sales over a similar period of time, but will not, at the end of that period, continue selling. Because by this time the product has become "old fashioned" to the traditionalist; the avant-garde group did not want it to begin with; and there is no other market receptive to it.

Mr. Coates suggests that the word "Concinnity" be used as descriptive of the type of design which the traditionalist intermediate market finds comfortable. When people receive visual sensory pleasure plus empathy, or a feeling for the product, and their backgrounds are compatible with the function of the product, then it has concinnity. Psychologically, if a thing is too boring it makes the person uneasy and, on the other hand, if the thing is too novel or "ugly" it makes the person uneasy also. Thus it behooves the designer to balance concinnity with novelty. Too much concinnity and the product will appeal only to the traditionalist market; too much novelty and the product will appeal only to the extreme avant-garde market, with resultant low sales level.

As a quick-look example, let us say that the passenger car profile today has a concinnity of:

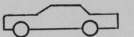

And that an avant-garde sport car profile might well be the profile of a Can Am racing car:

which would probably be unacceptable to most of the intermediate market and acceptable only to a few extreme avant-garde consumers such as 18 to 23-year-old men interested in racing cars.

So the designer of a sports car adds only a small bit of "novelty," such as a slightly undercut rear end and a small slope on the hood.

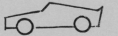 (Check out the 1975 Triumph 7 or the 1975 Fiat X-19)

We can assume now that it will sell reasonably well for about six months, with gradual increases in sales, because it has been accepted almost immediately by the majority of the intermediate-to-avant-garde market. Within one year, let's say, more traditionalists begin to accept this as comfortable and start to come into the market. They, in turn, support and probably increase sales, even though by now some of the avant-garde consumers are looking for a more novel design. This slight avant-garde design change in 1976 then becomes the concinnity of sport car buyers for 1978.

A conclusion from Coates's theory might be to base your mass-market design mainly upon the traditionalist intermediate point of view, with some definite elements of the avant garde incorporated into it, and your product will have a long sales life. Designing only for the straight traditionalist market will probably shorten the sales life of your product.

John Caldwell, product design instructor, Pasadena City College, offers the supplementary information below:

These assumptions may be true for the high-capital, complete-facility manufacturer who designs for a large, probably national, mass market. They may not be true for the low-capital, limited-shop manufacturer who mass produces items for local markets.

a. The local market is probably too small a sample. For example, it may well be very traditional oriented or very avant garde. So the smaller manufacturer usually must design specifically for his limited market.

b. The small plant can seldom beat the large operator in price. His costs per item are rarely less than those of a plant producing the same item in the millions. Thus again he must seek a much more specific market — either in a higher price range, a slightly different function, or perhaps a "hole" in the mass market left by the large manufacturer who didn't cover the entire desires or geographical area of the mass market.

DESIGNERS

Everything that man makes is designed. He designs things to the best of his ability whether it is in the area of physical function or visual appeal. Each discovery or invention is promptly modified, improved, and constantly redesigned with the idea of improving it. Designers have become a critical part of our fast-changing society.

In ancient times principles such as the lever, hammer, wheel and crossbow were probably progressive discoveries by unknown primitive brilliants.

Stone knife, fire by friction, rope, potters wheel, loom, needle, dyes, tanning hides, oil lamps, candles, mineral or graphite pencils, soap, paper, ink, gun powder, saw, nails, sundial, astrolabe, catapult, bellows, metal smelting, and tempering were primitive devices or discoveries that are the basis for almost everything we design today. Incidently (with the exception of gunpowder — please) how many of the above can you make from scratch? Probably one person in 5,000 has the knowhow to make a fire without matches, or steel, or magnifying glass. And of those perhaps one in twenty can actually produce a fire with his accumulation of sticks, raw-hide, and tinder. Try it some day.

In fact, it's rather a chilling thought that if you took the world's knowledge away from just one generation of children we would revert immediately to a caveman environment. And this with the added pleasant thought of probably having no chance to come back anywhere near as fast in machine inventions, because the metals of the world are long gone from the surface of the world. No "caveman" in, say, the year 2000 could happen upon a piece of pure iron, or happen to build his campfire on a copper deposit and get his insight from the funny molten slag left in his fire pit. Not on your life... He'd need, instead, a couple dozen bulldozers, some coring machinery, a small railroad, and dynamite, if he was going to get within a mile of any metals. Yes, aluminum is still plentiful, but you need one hell of an expensive and complicated mill to extract aluminum from its ore, even if you know where to find it and what it looks like.

There's a great book along these lines entitled "Earth Abides," by George Stewart, Random House, New York, NY. In it a very few people are left to cope with a city environment after a devastating plague kills off most of the population of the world. Besides the change in people's philosophy and outlook, it certainly pounds home the fact that we are all dependent on each other. And when too many of the specialists die, our culture dies with them. I feel it does have a message for the designer; "Are you certain this next design or robot you're working on is necessary?"

When you make everything too easy; when you give people everything; when you eliminate all physical work; when you take all decision making, creative activity and personal involvement away from a human being and protect him so completely he cannot fail at anything; you had better be careful. Because every invention and design in the past has been done with the idea of making life a little easier for that pestilence-ridden, flea-covered, slave-to-hard-work, insecure man of the Dark Ages. Which was great. But we've long passed that reason for design. Design today seems to be involved with too much trivial, insignificant, frosting on the cake—junk: remote controls for your TV set, automatic toothbrushes, electric knives

(Check the list on page 214. How many can YOU do?)

230

(maybe ok for a butcher, but what man can't, for Pete's sake, carve a roast with his own muscle? In fact, it's more fun if you do work at things a bit.), buzzers on your car dashboard to tell you when you're going over a set speed, wet deodorants, dry deodorants, roll on, wipe on, spray on, dab on — who the hell cares?

And, of course, there is no legislation against trivia (and there probably shouldn't be in a free society) or insignificance, or plain junk. If it sells ...wheee and away! And that, young designer, is where we're at today, 1977.

Maybe it's about time we, as designers and consumers both, started looking for a slightly different philosophy or reason for being. Suppose we designed for peoples' NEEDS, in the best way we could, but started taking it a little easy on designing every luxury item the affluent upper ten percent of the world's population can afford? We could conserve on the world's resources to a greater extent. We might even leave a few parks, some oil, and some metals for our grandchildren. And this type of restraint will obviously not happen overnight. It may not happen at all. But it CAN happen if designers start thinking about it. There is no idea like the idea whose time has come. And somebody has to make a start. Designers and engineers can pack a lot of wallop on world opinion, if they so desire. But THEY have to know where they're going, before they're able to convince anyone else. That's for sure. What do you think?

After the 1400's the spread of printing and communications in general seemed to generate a flood of machine design based on the discoveries of primitive and ancient man. The lever became the pliers and the steam shovel; the spring snare became the crossbow; glass became the lens in the telescope & microscope; the dugout and sailboat got a steam engine. Designers' names were attached to their invention or design, if it was an important discovery, and a certain prestige and economic benefits came the way of the designer through patents or copyrights. Today we have Revere Ware, Otis elevators, McCormick reapers, Colt pistols, Eames chairs, Olivetti typewriters and Land cameras.

TREE MONTS AGO
I CUDN'T SPEL DEEZINER.
NOW I ARE WON.

DESIGNER	WORK	YEAR
Salvino d'Armato	Spectacles	1300
Gutenberg	Movable type	1456
Leonardo DaVinci	Glider	1500
Hans Lippershey	Telescope	1608
Hargreaves	Spinning jenny	1764
Paul Revere	Coppersmithing	1765
James Watt	Steam engine	1769
George Washington	Agriculture	1770
Nicolas Cugnot	Automobile	1770
Tom Jefferson	Architecture	1770
Ben Franklin	Lightning rod	1775
	Bifocals	
John Fitch	Steamboat	1787
Eli Terry	Clock escapement	1792
Eli Whitney	Cotton gin	1793
	Interchangeable parts	
Oliver Evans	Steam auto	1804
Robert Fulton	Steamboat	1810
Michael Faraday	Electric motor	1821
Samuel Morse	Telegraph	1825
Thimonnier	Sewing machine	1830
Cyrus McCormick	Reaper	1831
Elias Howe	Sewing machine	1846
Samuel Colt	Revolver	1848
Lewis Temple	Whalers harpoon	1848
Elisha Otis	Power elevator	1857
Issac Singer	Sewing machine	1859
Dr. Richard Gatling	Machine gun	1862
Alfred Nobel	Dynamite	1866
Glidden & Sholes	Typewriter	1868
Emile Berliner	Gramaphone	1877
Tom Edison	Electric light	1880
	Motion pictures	
Nikola Tesla	A.C. motor	1882
Thomas Crapper	Flush toilet	1890
Marconi	Wireless telegraph	1890
Simon Lake	Submarine	1894
Charles Stienmetz	Electric motors	1896
Alexander Bell	Telephone	1896
Henry Ford	Assembly line	1906
Wright Brothers	Airplane	1906
G.W. Pickard	Crystal detector	1912
Eastman	Photographic film	
Waterman	Fountain pen	
Hunt	Safety pin	
Bostitch	Stapler	
Fermi, Einstein	Atomic energy	1943
Oppenheimer, Teller		
Bell Laboratory	Solar battery	1954
Land Corporation	Color polaroid	1970

ᙁᖳodern designers are not necessarily as well known as the old timers. One reason is undoubtedly the fact that most of the obvious breakthroughs or inventions have been done. It is a lot harder today to uncover a unique principle that hasn't already been worked to death. Instead, it usually takes a team of experts working under the auspices of a well-heeled corporation or government agency to come up with any important discovery. Also the modern designer tends to work on a greater variety of problems. With the improved communications, visual aids, information-retrieval systems, good trade publications, international seminars and the airplane, the contemporary designer has the world at his fingertips. He can often produce 10 times the gross product in his lifetime as that of his predecessors. Instead of working years on end on a trial and error basis, he can almost be assured of a successful solution if he has a clear concept. But to handle the complexities of design jobs in our fast-moving culture, he needs a good education. Just knowing how to fix a bicycle and the one-lung gasoline engine on the old farm doesn't prepare one to be a car designer any more.

Following is a list of more contemporary designers. Many of them have been involved in a multitude of design problems. Their combined areas of interest would encompass automotive design, furniture, hospital equipment, educational aids, computers, multiple-projection series, textiles, structural devices, typewriters, communication systems, transportation, packaging, office design, tableware, dishes, kitchen utensils, and design for the handicapped. In the administrative areas they have served on governmental advisory boards, as heads of national design bureaus, and as officers in industrial design societies as well as handling teaching assignments or running their own design offices.

Might be interesting to choose a name from the group and read up on his or her life: How they got educated, where they worked, where they lived and the different design solutions they were able to offer their clients make good educational background material for the budding designer.

Tomas Maldonaldo
Max Bill
David Pye
Marcello Nizzoli
Harold Van Doren
Walter Dorwin Teague
Henry Dreyfuss
Russell Wright
Buckminster Fuller
Raymond Loewy
Eliot Noyes
Andriano Olivetti
Ferdinand Porsche

Charles Eames
George Nelson
Peter Muller-Munk
Arthur Drexler
Gilbert Seldes
Robert Probst
Saul Bass
Edith Head
Kenji Ekuan (Japan)
John Reid (England)
Victor Papenak

Rodolfo Bonetto (Italy)
Frank Dudas (Canada)
Frank Height (England)
Sonja Bata (Canada)
Carl Aubock (Austria)
Henri Vienot (France)
Josine Des
Cressonnieres (Belgium)

CONTRACTS

There are a few suggestions below for the young designer when he begins to see the necessity for written contracts. Advice from experienced designers is usually couched in three words, "See a lawyer." However, not all lawyers are well versed in design contracts or patent rights. Those with excellent reputations may be too far away, too busy, or too expensive, so it behooves the young designer to at least acquaint himself with a few areas that have been "bones of contention" between clients and designers for years.

1. Keep the contract SIMPLE. Contracts are written to give you as complete protection as possible in court. However, there is a bit of philosophy here that you might consider. Usually as a young designer you cannot afford the fees and costs of a lengthy court case. Your objective should be, rather, as complete and simple an understanding between you and your client as possible. And, then, if things do go wrong, do the best you can to reach an agreement, but if all your attempts fail, forget it and go on designing for other clients.

    Keep in mind always that your forte, your strength, is originating. In a sense you are similar to a Research & Development (R & D) company. I'll always remember a staff meeting at an electronic R & D firm I was working for as a young man. The engineers in the plant (some of whom had stock in the company) were concerned over the fact that two of our best dynamic measuring instruments (a sensitive pressure pickup and an accelerometer) were being copied very blatantly and manufactured by a rival company, within 6 months after our company had developed and patented them.

    In answer to the question, "What are we going to do about it ?", the president of the company had this to say: "Well, I'm sure I feel as angry about it as you do. I've been mulling this thing over for about a week now, and I think I have an answer. Let's not forget that we started out as a development company. We have the fun of designing, engineering, placing components around the country for testing, and when we hit, we get orders and contracts for delivery almost immediately. I feel the industry is contracting for building almost two to three years ahead and moving so fast that no company can hope to exist by relying on pirating other products. As you all know we are in very good shape financially, we have a good reputation, and I think we have a pretty good esprit de corps in the plant, except for Joe." (Laughter. Joe was the toolroom check-out man who was always climbing everyone's back to return tools they took out.) "Anyhow, I, personally, would like to see us forget court battles and just stay so far ahead of those bastards they'll eat our dust. Besides, I have a lot more fun designing than I do arguing with lawyers, believe me."

Well, the upshot of it all was that most everyone agreed and, furthermore, felt a new surge of enthusiasm. Fifteen years later the company had grown to over 400 employees.

Moral: Know your strengths.

2. Try to state as specifically as possible just what you are going to do for your client:

    a. Are you going to sell him a SERVICE?

        Provide sketches and scale drawings? Working drawings? Will you be available for consultations on related products? Will you work with him regarding market surveys? Do you help engineer products in production? Will you reserve certain patent rights on patentable ideas?

    b. Or are you going to sell him an ITEM?

        Will you cause all rights to be vested in the client? Will your idea become his property? Will patent rights be his? Will he have the responsibility of marketing and sales? Will the client be responsible for all engineering? Will it be his obligation to make a valid attempt to sell the item he buys from you?

3. Set a few standards regarding ACCEPTANCE.

    a. After a new design is presented complete: It must be accepted or rejected, in writing, within thirty days? Maybe just checking a 'Yes' or 'No' box on the drawings would be sufficient. But you should establish some reasonable deadlines with the client so that he is not able to keep you "dangling" for months on end. You can't afford that kind of treatment. If he doesn't want it, maybe someone else does (unless you are tied to him for all ideas in that area).

    b. Once it is accepted: The client must start manufacturing the item within one year? Or all rights revert to the designer? Again, your livelihood often depends on royalties. And if your products don't get into production, you will starve. Are drawings and specifications then going to be returned to you? These are touchy subjects with many manufacturers. But it is certainly better to talk them out before either one of you is committed, and when you are both cool and working out an understanding, than to wait until a real tough emergency comes up with both parties red in the face and steam coming out of their ears.

4.   Make certain of the basis for PAYMENT.

   a.   You will receive royalties from, say, 3% to 5% on sales   (or flat fee, or hourly, or cost plus).

   Do you receive a monthly check? Is the percentage figured on items sold? Items delivered? or Items paid for? Items delivered is probably best for the lone designer. Then you get your check regardless of poor-paying customers. What about items delivered and then returned?

   Suppose an agent of your client in another state sells your items: Do you get royalties on those items also? Well, you better state in the contract that sales by "client or agent" are part of the ball game.

   Along with the check will the client furnish you with a monthly statement which lists the number of items sold that month? Will it be signed by a company official? Internal revenue spot checks often demand proof of this very nature. These precise records made at the time of transaction are a must for any tax preparation. Why not be prepared?

   b.   Suppose the client expects you to do some redesign work on an old standard item in the plant. Have you the right to reserve judgment as to whether you have the time or even the desire to work on such a project? Suppose you design a chair for your client: You get the royalties on the chair sales, all right, but you discover he is now manufacturing a footstool, a dining table, a coffee table and two end tables, using your chair design, fastener construction and detailing. These groups are selling like hot cakes, but you're not getting any royalties on the new items. You probably should work out something ahead of time regarding design "spread." Obviously your design ideas are making the group go. You should get royalties on all the items, as well as the chair. But most important, YOU should have been called in to design the group in the first place, to keep the group consistent. Your design ideas should no more be "pirated" by your client than by any competitor. Get this down as simply as possible in black and white.

   c.   Will royalties continue to be paid to your estate in case of your death? Might mention it in contract.

   d.   If royalties are not paid, will the client cease manufacture of your items at once? Or does he keep right on until you get a court order?

5. Consider a few AUDITING provisions.

    Will the client be willing to allow you or your agent (usually an auditing firm) access to the client's records — at least those pertinent to the designers products? And will they be available near your design office or in the corporation's offices in Birmingham, England? It might make a difference in an emergency. Will they be available once yearly? Or any time upon reasonable notice?

6. Consider CONFLICT-OF-INTEREST provisions.

    a. Most clients want to be assured that if you design for them you will not be designing any related products for another company. They may include in the contract an agreement that as long as they are manufacturing your designs and paying you royalties you may not design, say furniture, for any other company. This sounds reasonable enough. But let us suppose that the client's management changes, there is a hassle between you and the new management, and the new president decides that they are going to let your line die and gradually phase you out. Then for the next two or three years your royalties get less and less, until finally you are receiving only $30 per month royalties, and they are selling only two or three of your items per week. During all that time you may not, legally, according to the contract you signed 5 years ago, do any furniture designing for anyone else.

    b. It might be better not to accept such an agreement, but rather assure them that you are an ethical designer, you have always abided by the IDSA's code of ethics, and give them your word that you will not design any related products for any other company. Point out that if you did such a thing, your name would be Mudd* in the industry within weeks, and no one would want your designs anyhow.

    * Mudd was the doctor who set the leg of John Wilkes Booth without knowing he was aiding an assassin. He suffered persecution for years thereafter.

        You and the client may want to spell out more specifically exactly what "furniture" covers. Would they be concerned if you designed lamps for a lighting company? Theatre seating? Airplane lounges? Etc.

        Otherwise, you are completely free to design for others. Even then, it is always a good idea to let your clients know when you are taking on a new line before you commit yourself. Your old reliable clients, that have nursed you from a young unknown to a fairly well-known designer and pushed your products over the years to give you a darn good income, may very well be concerned over the fact that you are suddenly going to design for a company recently started by three design men who just split off from their firm 6 months ago. You may not have known any of this, and a quick commitment to the second company would certainly be considered a "conflict of interest." So use common sense and don't be afraid to consult with others occasionally. If you're a good designer you will make it on your designs and your character, not by treading on the fringes of your ethical commitments.

Pirating

It is very difficult to get an art design idea patented or copyrighted. And even if you do, the chance of protecting that right is rather slim. Pirating (stealing someone's design and reproducing it for profit) is rampant everywhere. The top design firms do not do it. They don't want that kind of reputation attached to their company. In fact, some designers often refuse to even look at young designers' portfolios for fear something they see might unconsciously be translated into one of their own designs.

The majority of older designers are not too concerned about someone stealing their designs. They go on the premise that they are good enough and strong enough to keep ahead of the copiers, get to the market first and have inventories ready before the parasites can figure out which item is "hot" and produce a cheaper, oftentimes shoddy, copy.

Fashions are probably a prime example of this. The first single "original" is produced and sold in some high-fashion salon in New York or Paris for a price well in the thousands. It is then usually seen at an opening or premier worn by a wealthy patron, copied within a few days and sent to London, Paris, New York and Los Angeles high-fashion shops. The prices during the first run are often in the hundreds of dollars. But within a few weeks the same design will probably be seen in almost every dress shop in the world for prices ranging from $15 to $50. Now there are many ethical dress designers who design for the lower-priced market and have reputations within the trade of being very creative and "saleable." Nevertheless, their designs are copied almost as fast as they get on the market too, in many instances, and so the game continues. Many manufacturers have the philosophy that, "Well, there's nothing new under the sun, anyhow, and if we don't get in on the hot items someone else will and run us out of business."

All in all then, it usually comes right down to what you think of yourself. Do you have the character to try to be original, to study the concept, research the approach and build a slightly better product? Or are you willing to spend your life copying others?

7. Patents or Design Ideas

Probably 85% of all design work is a remodification of existing designs.
However, you may come up with a patentable idea in the more limited area
of "new" products. If the design does qualify for a patent, the designer
often agrees to grant an "exclusive" LICENSE to his client in that particular
product area only. Suppose you have been allowed a patent on a suspension
theatre seat that telescopes into the back support to allow for sweeping out
three rows of seats at one run of a special broom? You could grant your
client, a theatre equipment company, an exclusive license for producing
such theatre seats, but you would still be able to grant other exclusive
licenses to, say, aircraft companies, railroads, or churches, etc., where
multiple seating does not compete with your client's special field.

This is usually better than "assigning" the complete patent to your client,
because the client can then grant licenses to other companies without you
getting a penny. And if the client goes bankrupt or discontinues your product, it is Billy-O-Hell getting a reassignment of patent back to yourself.

Suppose you don't get a patent, but your design idea is still quite novel and
unique. There are a few suggestions to help protect yourself. The three
typical payment "breaks" are: (1.) Research fees, (2.) Development and
presentation fees, (3.) Production follow-up fees. If possible, have an
understanding with your client on an OVERALL basis of payment. That is,
you may want it understood that merely paying you for your research and
presentation of ideas will not grant any rights to your design unless you are
in on the production design. Or if you ARE willing to release the design as
soon as the development and presentation are completed, then be prepared
to charge enough up to that point to remain solvent. There have been many
cases where the young designer made a presentation on a small retainer fee
and then watched the company pay him off, complete the design drawings
inplant and go into production design without consulting or paying the
designer.

This is not to say that all companies are dishonest. This is to say that both
you and your client should damn well understand WHAT you are selling,
WHEN you are releasing design ideas, and the PRICE you are going to get.
You are selling IDEAS, not sketches, and make certain the client knows
that BEFORE you get involved.

8. Deadlines

If a client wants to establish a specific deadline for say a completed presentation, a model or working prototype, or perhaps an exhibit booth, this is
fine. But suppose you have presented your sketches and preliminary proposal,
and the client sits on them for 6 weeks while he gets around to showing them
to his sales reps, art director, marketing manager, shop foreman, and mother-
in-law? One week before the set deadline he returns your proposal with
several changes and says, "Go." You will probably find that even if you
have your whole studio working on it day and night you can't make the
deadline. So you miss the deadline, the client is angry, you can't collect,
and your studio image is tarnished, because the client will tell everyone in
the trade how you goofed.

Suggestion: Any time a deadline is agreed upon put in writing that any preliminary work must be returned within, say, one week, or you cannot be re-
sponsible for meeting the deadline. Most clients will certainly see the necessity for this. In fact, many liaison men, such as the client's general manager,
prefer it, because it gives them a "contract lever" to force their subordinates
to make early decisions or get bypassed. So it keeps your image bright and
also keeps the client happy in the long run.

9.   Verbal Changes

Sometimes during a short meeting or a telephone conversation you and a
manager decide on some rather critical CHANGES. (Such items as delivery
schedule, types of materials to be used, who pays a new vendor, a design
change are some possibilities.) It is good business to type up a brief resumé
of the conversation and get the signatures of both the client and yourself on
it. You each then hold one copy.

Managers often say jokingly, "Don't you trust me, old boy?" and a good
comeback is, "I sure trust you, Weatheredstone, but I have had some experiences where people go on vacation, or leave the company, or even die, and
then weeks go by and past conversations are lost in the wind.   Let's get it
down in black and white, at least for our sakes, eh?"

And in some cases when you present the resumé of the conversation the
manager will say, "Wel-l-l, that isn't exactly what I meant ... You see
...., etc." Aha, so you see it wasn't as cut and dried as he and you thought
it was. Misunderstandings and wrong interpretations cause more hard feelings
in business than dishonesty ever did. Communicate clearly. To paraphrase
Kojak , "Writing (the law) isn't perfect, but it's a lot better than whatever
the hell is in second place."

DESIGN SOCIETIES   (Hdqtrs: 2 rue Paul Lauters, 1050 Brussels, Belgium)

ICSID, the International Council of Societies of Industrial Design, is a world-wide association of industrial design societies from all countries. It was created in 1957 by industrial designers who saw the need for international communication regarding new concepts and aspirations that were arising out of diverse economic and social structures in a variety of nations.

It is interesting that, perhaps with the possible exception of Finland, which established a designers'society in 1911, the majority of design societies were created between 1945 and 1969.

Artists in general do not lend themselves to organizational groups. The very nature of the beast has been that he remains independent, goes his own way, and thus comes up with creative ideas because he is NOT hemmed in by unions, guidelines, and committees breathing over his shoulder. So, although the professional industrial designer has existed for many years and has been recognized by corporations manufacturing a multitude of products, it is only in recent years, possibly the past twenty, that the profession is becoming a bit better known to the public. The increase in designer's curriculums in universities and community colleges has certainly been one indication of awareness by communities. The growth of the number of "World Fairs," "Centennials" and "Expositions" has also been instrumental in making people aware of design,and with it the fact that someone called a designer had to create the architecture, the images, the products and the environments.

A vocation starts becoming a profession when it begins to use its knowledge to assist mankind in general. To quote from the aims of the ICSID "...., and to initiate programmes which apply the industrial design process to the solution of problems which affect the material and psychological well-being of man." This quote states in a nutshell why we have had such a sudden growth of industrial design societies. It is not enough that the individual artist stays aloof, and goes his own way, if he wants to call himself a professional. These societies have become necessary; not to limit design, or legislate it; not to create restrictions and "juries"; but to expand communication among designers of all countries. And the single artist can no longer do that in today's world. If he is going to contribute toward conserving resources, preserving our ecology, and helping other people, he is definitely going to need organizations and men who are willing to give money or time in administrative capacities.

In passing then, the young designer who tends to "pooh-pooh" design societies ("I never was a joiner. What is there in it for me? All they ever do is sit around and talk or have some guy talk for two hours at a seminar. Not for me.") will probably be the first to crab because of government indifference to his central city redesign; the first to complain because the college doesn't give his design section as large a budget as the nursing section; and shake his head because students in European countries get scholarships to attend industrial design schools and he gets none. And after a trip to an underdeveloped section of his own country wonders why somebody doesn't DO something. Well, maybe, if he had attended that seminar, if he had considered putting part of his shoulder to the wheel, if he had been a member of his local design society and added to their slim budget with his dues... he might have at least understood the problem. And as he grew older he would be that better able to fight against the injustice of inner-city slums, to see that design curriculums receive their just share of the budget, and to DO something through his own design society to correct conditions in underdeveloped sections of his own state. Yes, young designer, if you really think you're in a profession you had better start keeping in touch with other professionals. And your membership in a professional society together with the annual dues is really a great start. Over the long, run you might be surprised at the personal pleasure you get from realizing that you have actually been a part of promoting good design the world around.

Following is a list of some of the industrial design societies in a variety of countries which belong to the ICSID. I have chosen only one from each country, although there are sometimes three or four different organizations in the same country. Membership varies from twenty or thirty members to the larger societies, such as Sweden's SS with 1845 members and England's SIAD with 6150 members.

The addresses are taken from the ICSID booklet 72/73. If you do happen to travel to any of these countries it might be worth your while to write a letter to one of the design societies and explain that you'd like to drop in to their office on or near a certain date. They are very helpful in giving you information concerning permanent design-exhibition centers, university design centers and other sources that every young designer should be aware of.

DESIGN SOCIETIES

Argentina*	CIDI	Maipu 171, Buenos Aires
Australia*	IDIA	151 Flinders St., Melbourne
Austria*	OE.I.F.	Salesianergasse 1, Vienna
Belgium*	ICSID	2 rue Paul Lauters, 1050 Brussels
	IBDID	72 Galerie Ravenstein, Brussels
Brazil	ABDI	Rua Santa Isabel 160, Sao Paulo
Bulgaria*	CEI	Rue Slavianska 8, Sofia
Canada	ACID	55 York St., Toronto, Ontario
Czechoslovakia*	UCA	Gottwaldovo nabrezi 250, Prague
Denmark*	IDD	Nikolaj Plads 9, Copenhagen
Finland	ORNAMO	Unioninkatu 30, Helsinki
France*	IEI	8 rue Jean Goujon, Paris
Germany*	VDID	Charlottenplatz 6 Hochhaus, Stuttgart
	AIF	Clara Zetkin Strasse 28, Berlin
Gr. Britain*	SIAD	12 Carlton House Terrace, London S.W.1
Hong Kong	HKIDC	21 Ma Rau Wei Rd., Hong Kong
Hungary	AHFA	Vorosmarty ter 1. x/1029, Budapest V
India	NID	Sanskar Kendra-Paldi, Ahmedabad 7

Ireland*	KDW	Kilkenny Design Workshops, Kilkenny
Israel	IDC	51 Petah-Tikva Rd., Tel Aviv
	IIDA	11 Kikar Malchei Israel, Tel Aviv
Italy	ADI	Via Clerici 5, Milan
Japan*	JIDA	4-1 Hamamatsu-cho, Tokyo 105
Korea	KDPC	128 Yunkung-dong Chongro-ku, Seoul
Mexico	AMD	Insurgentes Sur 1443, Mexico City
Netherlands	SIV	Beurs Damrak 62a, Amsterdam
New Zealand	NZSID	P.O. Box 3432, Auckland
Norway*	ID.NDI	Drammensveien 40, Oslo 2
Poland	SPFP	Skr. Poczt 14, Warsaw 12
So. Africa*	SIASA	Plaut Interiors, Foreshore, Capetown
Spain*	ADI/FAD	Calle Brusi 39-43, Barcelona 6
Sweden*	SSID	Malmskillnadsgatan 48A, Stockholm
Switzerland	VSID	Alte Landstrasse 415, Mannedorf
Taiwan	CPC	62 Sining South Rd., Taipei
U.S.A.	IDSA	1750 Old Meadow Rd., McLean, VA 22101
U.S.S.R.*	VNIITE	Research Institute of Industrial Design, Moscow I-223
Yugoslavia	SULUPUDJ	c/o Dlos, Titova 21, Ljubljana

*Have Industrial Design Exhibition Centers. With one or two exceptions the
Design Centers are in the same city as the design society headquarters.
These centers vary in size from 200 to 1200 square meters or more and exhibit
products of the country's industrial designers,which include capital as well as
consumer goods. Attendance varies anywhere from 200 to 4000 visitors per day.
The objectives are design improvement, promotion for export and domestic sales,
as well as educational awareness. Financial support comes usually from the
national budget and partly from space rentals.

Altho many designers use cold-type (wax-backed or transfer) letters for their headings on presentations, students cannot always afford them. Most art stores give out the free cold-type catalogs supplied by the manufacturers, and several of these on the lab bookshelf can provide "lifts."

Draw a line on a sheet of tracing paper. Choose your alphabet and place the tracing paper over it, keeping the line exactly even with the base of each letter as you outline trace your heading. Keep your letters almost touching each other for spacing. If your presentation board is a light value, darken the back of your tracing with a 4-H pencil. (Softer pencils are too greasy and make for rough paint edges.) Trace your heading on the board and fill in the letters with water-base paint & a number 6 sable. If your board is a dark value, use white chalk on the back of your tracing.

The partial alphabets below will give you a practice sheet.

ABCDEGIKMSUY
1234567890 abefk

ABCDEGHKMNOPRSTY, 2345

ABCDEGHKMNOPRST, 234568 ad

ABCDEGKNS. 2345

ABCDEGHKMNPQRSTW
(Too small for lift. Could be enlarged in "Lucy.") 12345678

ABCDEGHKLMNS
rstw abcdefghiklm

244

ALPHABETS

Graphic Display

The designer is constantly faced with the problem of communicating
information to the user of his product via letters, figures or symbols.
These can be printed, stamped, embossed, etched or silk screened
on a variety of surfaces.

If the product itself has to produce the information by means of
read-out panels, display lights, cathode tube grids, or what have
you, then it is essential that the simplest possible alphabet be used.
The greater the complexity of the letterform, the greater the
number of letters necessary to do the job, then the greater is the
problem of the computer or keyboard operator. Through the ages
the English alphabet has been added to willy-nilly. At the present
time there are probably 7 or 8 letters that are absolutely unnecessary, and many of the letter formations could easily be modified to
make identification simpler and faster. Let's take a quick look at
some future possibilities.

Many years ago the Whitman Chocolate Co. started marketing a box
of chocolates with a variety of center flavors. It was called the
 Sampler , so you could sample the various flavors. This name
was embroidered on the box top using a square grid pattern as a
base for the letters, and the name, Sampler, then looked as if it
were made from a set of little squares. Hence the name, Sampler
Readout , was given to similar letterforms on computer displays
today. This grid system or pattern of arranging lights on the
display board is a way to display letters and numbers on a variety
of equipment.

A common matrix used is a 5 x 7-dot panel which contains thirty-five
lights or "bits" to simulate our letterforms. This format is often seen
in electric signs and scoreboards on baseball diamonds.(See examples.)

However, this matrix has the capacity to produce millions of characters, which is way beyond what we need to communicate ... but the
complex way our Roman letters are shaped dictates that we need a
complex board to simulate all the extra letters and traditional
shapes of our mediaeval alphabet.

In the January-February, 1974, issue of "Industrial Design" magazine there is a very important article titled,"Simplifying the ABC's,"
by a Design Planning Group of which Jay Doblin is "Concept"
member. Every young product designer should attempt to find this
article and read it. It gives, in essence, a clear, concise picture
of alphabet use from the stylus and quill pen to modern machines.
It plugs the idea of hastening the simplification of alphabet forms
by use of a seven-bit rectangle and eliminating some of the unnecessary letters in our alphabet such as U, V, W, X, Y, Z, Q, J, and K.

245

G

AMBIGUITIES:

GIRAFFE
GROUND
GREAT
GENERAL
      JENERAL ?
GACK
      JACK ?
GEORGE
      JEORJE ?
GUDGE
      JUDGE ?

SIMPLIFY:

KANGAROO
   CANGAROO
KEEP - CEEP
QUARTER
   COARTER
QUEEN
   COEEN
VACCINE
   FACSEEN
WRITE - RITE
XEROX
   SEROCS
BERRY - BERRI
YARD - IARD
YACHT - IAT
TOMORRO (W)

As you probably realize already, there are many irregularities and much downright stupidity in alphabet usage in the English language. George Bernard Shaw once pointed out that the word, fish, could be spelled "ghoti". The gh from enough; the o from women; the ti from nation. Another example is the pronunciation of the words cough, rough, through, and though: cof, ruf, throo, and tho. Pity the poor translating machine that has to analyze those phonetic blunders. And hasn't the school child (another translating machine) been even more confused?

A B C D E F G H I J̸ K L M N O P Ǫ R S T Ʉ Ѵ̸ Ѡ̸ X̸ Y̸ Z̸
     K   V J   Y   U   Z
     Q   W
     X

The letters J, K, Q, U, V, W, X, Y, Z are really unnecessary to pronunciation.

The u sound is indicated by oo.
The w sound is already indicated by the o in such words as one.
The z sound is close enough to the s sound.
The q is a hard c, as is k, etc.

Considering each bar as a "bit," the rectangle below would be 7 bits, used then as a readout panel with the resulting alphabet of:

8   A b C d E F G H I L Ⴖ ∩ O P r S Γ
    A B C D E F G H I L M N O P R S T

1 2 3 4 5 6 7 8 9 0
1 2 3 4 5 6 7 8 9 0

A more extreme modification would be possible by reducing the alphabet to 15 letters and letting the numbers share characters with the letters. This is pretty far out for now, but all the information necessary could then be carried by a four-bit symbol; and our alphabet would look something like the following: (What are the last two letters you would eliminate to get the number of characters down to fifteen?)

A	B	C	D	E	F	G	H	I	L	M	N	O	P	R	S	T
	6	2	3			7		8	1	4	9		0		5	

C or G ?
B or P ?
A or E ?

246

Would using a 5-bit format help matters? Could we recognize the numbers easier if they weren't exactly similar to letterforms?

Bꓶ AꓳꓒꓸꓳEFꓞHILꓵꓵ Oꓵꓳꓲꓶ

A B C D E F G H I L M N O P R S T

I ꓵꓳꓕꓵꓶꓷBꓯꓳ

1 2 3 4 5 6 7 8 9 0

It is definitely feasible to redesign our alphabet. We have the
solution. Only inertia and tradition will keep us using old-fashioned
methods. Machines such as typewriters, linotypes, TV , perfora-
tors, ticker-tape symbols, etc., have changed our reading habits so
that our speed of comprehension has outrun an alphabet designed for
the quill pen. From the hieroglyphics to our present letterforms,
simplicity has always been the trend. This is probably the first era
in history that man can really plan his language by redesigning the
alphabet ahead of time. If we can introduce the metric system into
our schools by starting in the lower grades, there is no reason a
simplified alphabet couldn't be introduced in the same way. Why
wait two centuries for tradition to change visually and phonetically
what we know is needed now and is coming anyhow?

With all this information then: Let us suppose you were a designer
living in the year 2078. Sketch up a typewriter that uses the 4-bit
alphabet. What would be the minimum possible number of keys
necessary on the keyboard? Not necessarily printing keys, now,
but KEYBOARD keys — simple, eh?

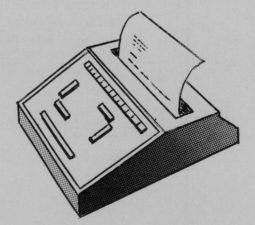

WEIGHTS & MEASURES

All the industrialized nations of the world use the metric system, except the United States and Canada. The metric base of 10 is much easier to understand, easier to use and relates to world measurements such as distance between meridians at the equator. Some day it will become the principal standard in the U.S. also, so it behooves the young designer to become acquainted with it. Designers and engineers use decimal, metric, foot and pound equivalency charts. Mechanics need two sets of wrenches, metric and regular. Speedometers often have both kilometer and mile indications. The time wasted interfacing designs and looking up equivalents between the two standards must be phenomenal.

In the meantime, however, some of the relationships and rough equivalencies between the metric and our foot and pound system (based on 12 ?) are listed below: Good luck !

WEIGHTS AND MEASURES

Circle Area $= $ dia.^{2} x .7854

radius^2 x 3.1416

Circle Circum. $=$ dia. x 3.1416
Triangle Area $=$ base x 1/2 altitude
Sphere Surface $=$ dia.^{2} x 3.1416
Sphere Volume $=$ dia.^{3} x .5236
Cone Volume $=$ base area x 1/3 altitude
Pyramid Volume $=$ base area x 1/3 altitude

American Standard
Wire & Metal Gauge

	Gauge	Inches
▬	8	.128
▬	10	.102
▬	12	.081
▬	14	.064
▬	16	.051
▬	18	.040
▬	20	.032

Household Measure

60 drops	=	1 teaspoon
3 tsps	=	1 tablespoon
16 tbs	=	1 cup
2 cups	=	1 pint

Caliber = Hundredths of an inch

Mariners Measure

1 fathom $=$ 6.08 ft.
1 nautical mile $=$ 6080 ft.
1 nautical mile $=$ 1° longitude at equator
1 knot = speed of 1 nautical mi. per hr.
Water pressure in lbs. per sq. in. = column ht. in ft. x .434

Temperature Conversion

F Fahrenheit = Degrees Centigrade x 9/5 + 32
C Centigrade = (Deg. Fahrenheit – 32) x 5/9
K Kelvin $=$ Degrees Centigrade + 273.16
R Rankine $=$ Degrees Fahrenheit + 459.69

* These last two use absolute zero on the F or C scale.

(Celsius was the original centigrade scale inverted.)

METRIC MEASURE

	km	Kilometer	=	1000. meters
(dk)	da	Dekameter	=	10. meters
	m	Meter	=	1. meter
(dcm)	d	Decimeter	=	.1 meter
	cm	Centimeter	=	.01 meter
	mm	Millimeter	=	.001 meter

1 Meter	=	1,000,000 microns
1 Meter	=	1,000 millimeters
10 mm	=	1 centimeter
10 dcm	=	1 meter

1 inch	=	2.54 centimeters
1 foot	=	3.048 decimeters
1 yard	=	.9144 meter
1 mile	=	1.6093 kilometer

1 sq. in.	=	6.452 sq. cm.
1 sq. ft.	=	9.2903 sq. dm.
1 sq. yd.	=	.8361 sq. m.
1 sq. mi.	=	2.59 sq. km

1 cu. in.	=	16.39 cu. cm.
1 cu. ft.	=	.0283 cu. m.
1 cu. yd.	=	.7645 cu. m.

1 mm.	=	.03937 in.
1 cm.	=	.3937 in.
1 dcm.	=	3.937 in. = .328 ft.
1 m.	=	39.37 in. = 1.0936 yds.
1 km.	=	.62137 mi.

1 sq. cm.	=	.1550 sq. in.
1 sq. dcm.	=	.1076 sq. ft.
1 sq. m.	=	1.196 sq. yd.
1 sq. km.	=	.386 sq. mi.

1 cu. cm.	=	.061 cu. in.
1 cu dcm.	=	61.022 cu. in.
1 cu m.	=	1.3 cu yds.

1 Hectare	=	10,000 sq. m.
1 Hectare	=	2.471 acres

Weights

1 grain	=	.0648 gram
1 gram	=	wt. of 1 cu. cm. water
1 gram	=	.03527 ounce
1 kilogram	=	1000 grams
1 kilogram (kilo)	=	2.2046 lbs.
1 metric ton	=	1000 kilograms
1 metric ton	=	2205 lbs.
1 ounce	=	28.35 grams
1 pound	=	.4536 kilograms
1 liter water	=	2.2046 lbs.
1 cu. ft. water	=	62.32 lbs.

Liquid Measure

1 pint	=	16 fluid ounces
2 pint	=	1 quart
4 qts.	=	1 gallon
1 gal.	=	231 cu. in.
1 Br. gal.	=	1.2 U.S. gal.
1 cu. ft.	=	7.48 gal.
1 cu. ft. water	=	62.32 lbs.

Equivalents (Liquid)

1 liter	=	1 cu. decimeter
1 liter water	=	1 kilogram
1 liter	=	1.0567 U.S. quart
1 U.S. qt.	=	.9464 liter
1 Br. qt.	=	1.136 liter
1 U.S. gal.	=	3.785 liter
1 Br. gal.	=	4.544 liter

Dry Measure

1 quart	=	67.2 cu. in.
1 quart	=	2 pints
8 quarts	=	1 peck
4 pecks	=	1 bushel
1 U.S. bu.	=	2150.4 cu. in.

Equivalents (Dry)

1 liter	=	.908 U.S. dry qt.
1 liter	=	.880 Br. dry qt.
1 quart	=	1.101 liters
1 Br. bu.	=	1.032 U.S. bu.
1 Br. bu.	=	36.35 liters
1 U.S. bu.	=	35.24 liters

DESIGNERS CODE OF PROFESSIONAL CONDUCT

The code of conduct which designers subscribe to has come down through the centuries from ancient artisans to craftsmen in the mediaeval guild into its present day form. It is exemplified in the code published by the International Council of Societies of Industrial Design (ICSID), Belgium, and modified versions may be found in many national design societies the world over.

The following is a brief outline of code principles to acquaint the student designer with the accepted basis of professional ethics.

1. Community

    Designer has an obligation to further the social, ecological, and esthetic standards of his community.

2. Client

    a. Designer shall not work on outside assignments which are in competition with client's products without client's permission.

    b. Designer and his staff shall treat all knowledge of his client's business as confidential. This includes client's intentions, business methods and organization, as well as his production processes and products.

3. Other Designers

    a. A designer must not try to supplant another designer or compete with another designer by deliberate reduction of fees or other inducements.

    b. Members shall not accept an assignment on which another designer has been working except with the agreement of the other designer or unless the previous committment has been properly terminated.

    c. A designer must not knowingly plagiarise any design.

4. Fees

    a. A designer should not do any work at the request of a client without payment of an appropriate fee. (However, he may do so for nonprofit or charitable organizations.)

    b. He should not retain any discounts, commissions or allowances of any sort from suppliers or contractors.

    c. If he owns stock in, or is otherwise involved with, a company that will benefit from his recommendations to a client, he must notify his client of this fact.

5. Publicity

    a. A member may not take advertising space or time in any media in which to advertise his professional services.

    b. However, the designer may issue to news media any illustrations, factual descriptions of his work and biographical material, provided he does not pay for such insertions.

    c. The designer must not release for publication to the press or otherwise any information about work he is doing for a client unless client gives consent.

    d. He may advertise a vacancy in his organization in the classified columns of appropriate publications.

6. Approaching Potential Clients

    a. The designer may use a letter addressed to an individual in a company. It can include a resumé of designer's experience, qualifications, and former clients but is not usually accompanied by any samples, photographs, or enclosures of any sort unless later requested by the company. (Such unsolicited letters may be sent only once in a twelve-month period.)

        Sending an unsolicited letter to a company already using another designer in the same field may be regarded as evidence of an attempt to supplant and therefore be considered grounds for expulsion from the society.

    b. He may present specimens of work by specific invitation from clients, or when applying for a salaried appointment. He may also present specimens in those areas where by custom an open invitation to do so already exists. For example, advertising agencies, publishers, textile and wallpaper manufacturers sometimes accept designs for consideration.

    c. May insert an advertisement for a salaried position in classified columns of appropriate magazines or newspapers.

7. Competitions

    Designers may take part in any open or limited competition whose terms are approved by the professional society of which he is a member.

Realize that there are many designers' organizations in the world, and that this code may vary from society to society, even within a country. If one does design work in a foreign country, it is an excellent idea to check on the code of that particular area or country.

ART SCHOOLS

The art schools listed below vary from those having a few core classes in product design or 3-dimensional design to those having a complete four-year industrial design curriculum with a variety of related subjects. Not listed here are the majority of large state universities, which usually have some type of product design program also.

ALABAMA

Auburn University
School of Architecture and Arts
Auburn, Alabama 36830

ALASKA

Anchorage Community College
1700 West Hillcrest Drive
Anchorage, Alaska 99502

ARIZONA

Phoenix College
1202 West Thomas Road
Phoenix, Arizona 85013

Industrial Design Section
Dept. of Technology
Arizona State University
Tempe, Arizona 85281

ARKANSAS

Arkansas Polytechnic College
Russellville
Arkansas 72801

CALIFORNIA

Academy of Art
740 Taylor Street
San Francisco, Calif. 94102

Department of Industrial Design
Art Center College of Design
1700 Lida Street
Pasadena, CA 91103

Dept. of Art
Stanford University
Palo Alto, Calif. 94305

CALIFORNIA

Claremont Graduate School
Claremont
California 91711

East Los Angeles College
5357 East Brooklyn Avenue
Los Angeles, Calif. 90022

Industrial Design Section
School of Fine Arts
University of California, Long Beach
Long Beach, Calif. 90801

El Camino College
Crenshaw and Redondo Beach Blvd.
Torrance, Calif. 90506

L. A. Valley College
5800 Fulton Avenue
Van Nuys, Calif. 91401

Pasadena City College
1570 East Colorado Blvd.
Pasadena, Calif. 91106

Rio Hondo Junior College
3600 Workman Mill Road
Whittier, Calif. 90601

San Jose State College
125 South Seventh Street
San Jose, Calif. 95114

Dept. of Industrial Design
California Institute of the Arts
Valencia, California 91355

San Diego State College
San Diego
California 92115

Los Angeles Trade-Technical College
400 West Washington
Los Angeles, CA 90015

CALIFORNIA

University of Calif. at Los Angeles
405 Hilgard Avenue
Los Angeles, Calif. 90024

Woodbury College
1027 Wilshire Blvd.
Los Angeles, Calif. 90017

San Fernando Valley State College
18111 Nordhoff Street
Northridge, Calif. 91324

San Francisco Academy of Art
625 Sutter
San Francisco, Calif. 94102

CONNECTICUTT

College of Engineering
University of Bridgeport
Bridgeport, Connecticut 06602

DELAWARE

Design Section, Art Department
University of Delaware
Newark, N.J. 19711

FLORIDA

Embry Riddle Aeronautical Institute
Box 2411
Daytona Beach, Florida 32015

GEORGIA

Berry College
Mt. Berry
Georgia 30149

Shorter College
Rome
Georgia 30161

Georgia Institute of Technology
225 North Ave N.W.
Atlanta, Georgia 30332

ILLINOIS

University of Illinois
College of Applied Arts
Champaign, Illinois 61820

School of Design & Architecture
Southern Illinois University
Carbondale, Ill. 62901

Industrial Design Section
Dept. of Art
Northern Illinois University
DeKalb, Illinois 60603

Industrial Design Section
Illinois Institute of Technology
Institute of Design
3360 South State Street
Chicago, Illinois 60616

Industrial Design Section
Dept. of Design
College of Architecture & Art
Box 4348
Chicago, Illinois 60680

Industrial Design Section
Transportation Center
Northwestern University
Evanston, Illinois 60201

Illinois Wesleyan University
210 East University
Bloomington, Illinois 61701

INDIANA

Indiana University
Department of Fine Arts
Bloomington, Indiana 47401

Industrial Design Section
Department of Art
University of Notre Dame
Notre Dame, Indiana 46556

Industrial Design Section
Dept. of Creative Arts
Purdue University
Layfayette, Indiana 47907

KANSAS

Kansas City State University
Manhattan
Kansas 66502

Kansas State College of Pittsburgh
1700 South Broadway
Pittsburgh, Kansas 66762

Industrial Design Section
School of Fine Arts
University of Kansas
Lawrence, Kansas 66044

MAINE

Colby College
Waterville
Maine 04901

Haystack Mt. School of Crafts
Deer Isle
Maine 04627

MARYLAND

Maryland State College
Princess Anne
Maryland 21853

MASSACHUSETTS

Brandeis University at South Bend
Waltham
Massachusetts 02154

Tufts University
Medford
Massachusetts 02155

Williams College
Art Department
Williamstown, Massachusetts

Massachusetts College of Art
Brookline and Longwood Ave.
Boston, Mass. 02215

Product Design Section
Center for Advanced Visual Studies
Massachusetts Institute of Technology
Cambridge, Mass. 02138

MICHIGAN

Andrews University
Art Department
Berrien Springs, Michigan 49104

Cranbrook Academy of Art
500 Lone Pine Road
Bloomfield Hills, Michigan 48013

Delta College
University Center
Michigan 48710

Northern Michigan University
Department of Visual Art
Marquette, Michigan 49855

Dept. of Industrial Design
University of Michigan
Ann Arbor, Michigan 48104

Industrial Design Section
Kresge Art Center
Michigan State University
East Lansing, Michigan 48823

Industrial Design Section
College of Art & Design
245 E. Kirby
Detroit, Michigan 48226

MINNESOTA

Minneapolis School of Art
200 East 25th Street
Minneapolis, Minnesota 55405

Environmental Design Section
School of Associated Arts
344 Summit Ave.
St. Paul, Minn. 51005

MISSISSIPPI

Meridian Junior College
Meridian
Mississippi 39301

MISSOURI

Design Section
Kansas City Art Institute
4415 Warwick Blvd.
Kansas City, Missouri 64105

MONTANA

Montana State University
Bozeman
Montana 59715

NEW JERSEY

Newark State College
Union
New Jersey 07083

Newark School of Fine and Industrial Art
550 High Street
Newark, New Jersey 07102

School of Architecture & Urban Planning
Princeton University
Princeton, New Jersey 07112

NEW YORK

Bard College
Division of Art
Annandale-on-Hudson, New York 12504

College of New Rochelle
Department of Fine Arts
Castle Pl., New Rochelle
New York 10805

New School for Social Research
66 West 12th
New York, New York 10011

New York University
Division of Continuing Education
1 Washington Square North
New York, New York 10003

St. Johns University
Grand Central and Utopia Parkway
Jamaica, New York 11432

NEW YORK

State University College
13000 Elmwood Avenue
Buffalo, New York 14222

State University
New Paltz
New York 12561

State University of New York
1400 Washington Avenue
Albany, New York 12203

State University
Fredonia
New York 14063

State University
Stoney Brook
New York 11790

City College of City of New York
Convent Avenue at 138th Street
New York, New York 10031

University of Rochester
Department of Art
Rush Rhees Library
Rochester, New York 14627

Vassar College
Department of Art
Raymond Avenue
Poughkeepsie, New York 12601

Sarah Lawrence College
Admissions Office
Bronxville, New York 10708

Skidmore College
Saratoga Springs
New York 12866

Center for Environmental Studies
Cornell University
Ithaca, NY 14850

Alfred University
Alfred, NY 12208

NEW YORK

Dept. of Industrial Design
School of Art
Syracuse University
Syracuse, NY 13210

Parsons School of Design
410 E. 54th Street
New York, NY 10022

Dept. of Industrial Design
Pratt Institute of Art
215 Ryerson Street
Brooklyn, NY 11205

School of Visual Arts
209 East 23rd Street
New York, New York 10010

NORTH CAROLINA

East Caroline College
School of Art
Box 2704
Greenville, North Carolina 27834

North Carolina State University
School of Design
Raleigh, North Carolina 27607

Western Carolina College
Collowhee
North Carolina 28723

School of Design
Western Carolina University
Cullowhee, NC 28723

Product Design Section
School of Design
University of North Carolina
Raleigh, NC 27607

OHIO

Dept. of Industrial Design
College of Design, Arch. & Art
University of Cincinnati
Cincinnati, Ohio 45221

Department of Industrial Design
The Dayton Art Institute
Forest & Riverview Avenues
Dayton, Ohio 45401

Industrial Design Section
Cleveland Institute of Art
11141 East Blvd.
Cleveland, Ohio 44106

Dept. of Industrial Design
Columbus College of Art & Design
486 Hutton Place
Columbus, Ohio 43215

Dept. of Industrial Design
Columbus College of Art & Design
486 Hutton Place
Columbus, Ohio 43215

Ohio State University School of Art
126 North Oval Drive
Columbus, Ohio 43210

Ohio University
School of Art
Athens, Ohio 45701

OKLAHOMA

Oklahoma Christian College
South Eastern and Memorial Road
Oklahoma City, Oklahoma 73111

PENNSYLVANIA

Academy of the Arts
107 6th Street
Pittsburgh, Pennsylvania 15222

Art Institute of Pittsburgh
635 Smithfield Street
Pittsburgh, Pennsylvania 15222

Carnegie Institute of Technology
College of Fine Arts
Pittsburgh, Pennsylvania 15213

Kutztown State College
Kutztown
Pennsylvania 19530

Philadelphia College of Art
Broad and Pine Streets
Philadelphia, Pennsylvania 19102

RHODE ISLAND

Coker College
Hartsville
Rhode Island 02913

Research and Design Institute
Providence, R. I. 02903

Industrial Design Section
Rhode Island School of Design
2 College Street
Providence, R. I. 02903

SOUTH DAKOTA

Augustana College
29th and S
Sioux Falls, South Dakota 57102

Northern State College
Aberdeen
South Dakota 57401

TEXAS

Odessa College
Post Office Box 3752
Odessa, Texas 79760

University of Houston
Art Department
Cullen Boulevard
Houston, Texas 77055

UTAH

University of Utah
College of Fine Arts
Salt Lake City, Utah 84112

VIRGINIA

Hampton Institute
Hampton
Virginia 23666

Richmond Professional Institute
School of Art
Richmond, Virginia 23220

WASHINGTON

Burnley School of Prof. Art
905 East Pine Street
Seattle, Washington 98122

Industrial Design Section
Dept. of Art
University of Washington
Seattle, Washington 98105

WISCONSIN

Marian College of Fond du Lac
45 National Avenue
Fond du Lac, Wisconsin 54935

Milwaukee Institute of Technology
1015 North 6th Street
Milwaukee, Wisconsin 53203

Stout State University
Menomonie, Wisconsin 54751

Layton School of Art
4650 N. Port Washington Road
Milwaukee, Wisconsin 53212

THE FUTURE ?

One of the most creative design ideas regarding world problems was proposed by Buckminster Fuller at the Edwardsville Campus of the Southern Illinois University about 1972. He called it "World Game". Mr. Fuller's basic idea was that there is a definite concrete alternative to waste, pollution and poor distribution of world resources; and that we have the knowhow and resources to solve such world-wide problems right now.

His proposal was to establish a computer complex with large-scale readout panels showing ALL world statistics. The nations of the world would help submit as much pertinent information as possible, and then the staff at World Game center would inventory, classify and program every world, regional and local resource (income) and every operation or need (expenses) in computers. The information called for from the computer's memory banks could then be plotted and used to help control the destiny of our own spaceship, Planet Earth, in a saner, more logical manner than we now seem to be doing.

The world resources inventory would show such items as:

1. Location of all the peoples on earth and how they move
2. Climate and weather patterns
3. Food crops, animals and sea life
4. Fuel resources: coal, oil, water, atomic, tides, & thermal
5. Metals: iron, aluminum, zinc, copper, etc.
6. Fibres: cotton, flax, wool, etc.
7. Transportation: ships, railroads, aircraft, trucks, pipe lines, etc.

The world operation or needs inventory would include:

1. World-wide food-consumption patterns
2. World product consumption: housing, clothing, furniture, appliances, medical, etc.
3. Energy consumption: electricity, gasoline, heating, lighting, etc.
4. Communications: radio, TV channels, telephone, satellites, etc.

There are many factors involved: Regarding energy, the world needs lighting on one side of the earth when the other side doesn't need it. And with new inventions and boosters, electrical energy can be economically transported for thousands of miles, and with it comes the possibility for a world-wide electrical network. This would mean that power could be used when needed and where needed regardless of source location. Day-night, seasonal hookups could bring per capita kilowatt levels up all over the world, instead of merely cutting off dynamos during local slack periods. Also with round-the-world networks, hydro-electric power might finally be feasible from tide movements.

Some computer results have already pointed up the ridiculousness of unplanned logistics. For example, in 1967, Asia exported and imported the SAME amount of rice. Figures show also that in too many cases up to 80 percent of food tonnage can be lost in transport, processing, storage and spoilage, merely because of poor planning or usually no international logistic planning at all.

A multi-million dollar computer center is probably the only way all these complex world inputs and outputs can be handled logically. Successful corporations already operate on these same basic principles: knowing plant inventories; knowing supply sources and prices; knowing plant capability in terms of both facilities and labor; knowing present market and estimated future markets; etc. Corporations are constantly making studies years ahead and equating the results with their current practices. There is no reason we can't operate our world on similar, business-like, logical principles instead of wasting it away by continuing to ignore tomorrow. No wonder experts tell us we are on our way to planet bankruptcy.

The center could eventually expect to parallel or relate to newscasts by sending out evaluations and possible solutions regarding world economics and design over the intermedia world network of TV, radio, newspapers and magazines. The physical evaluations and proposed alternate solutions could be a major source of world information BEFORE the fact, much as present newscasts are ordinarily AFTER the fact.

Well, there isn't near enough room here to even start to explain all the possibilities that "Bucky" envisions for the future from World Game. But it is a very exciting practical possibility of ending pollution, reducing starvation and possibly alleviating cold wars. Obviously there are philosophical and social attitudes that need changing. The jealousies between nations are not going to disappear overnight. The present "Them or us." attitude will need considerable softening into a "Them AND us." attitude before international cooperation would allow such an idea to flower.

The present location of World Game is at the University City Science Center in Philadelphia. The staff is hoping to provide enough intangible as well as tangible information from world sources to modify business decisions instantaneously on a day-to-day basis. The entire idea has overtones of being the basis for the next era of human development. The industrial revolution ushered in the Atomic-Space Age with such a blast that man merely DID in open-mouthed amazement without realizing all the consequences. World Game may well be the start of The Information Age, where we control our planet's environment through world-wide reason and DESIGN.

How would you like to work on a design job of this size? Well, many of you students are going to have the desire, the drive and the knowhow to continue in Mr. Fuller's footsteps someday. Keep it in mind. The world needs you.

See also "Project Link," a computer model of world economics with 3000 equations, in "Fortune" magazine, March, 1975, page 156.

INDEX

Z